The Art of Paper

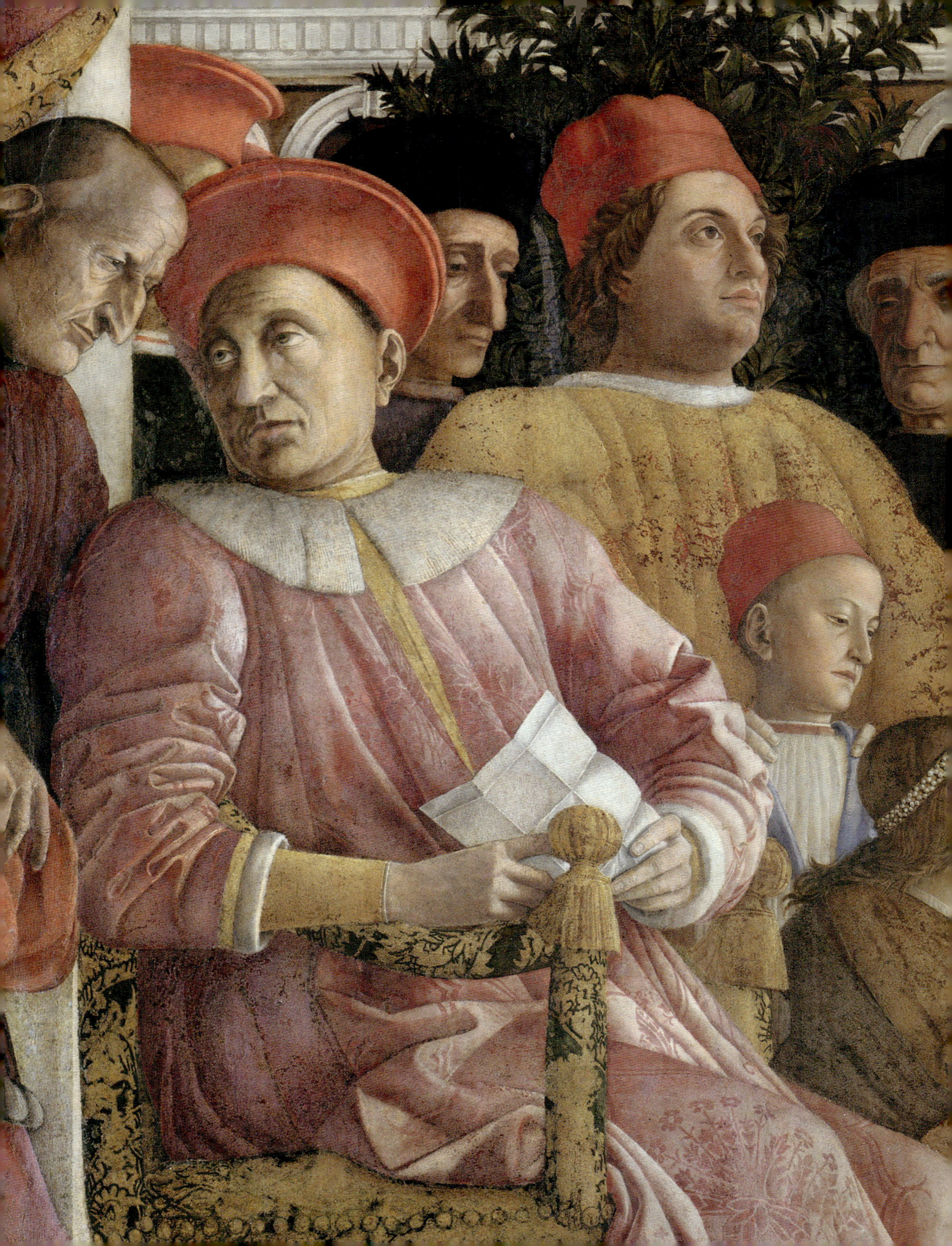

The Art of Paper

From the Holy Land to the Americas

Caroline Fowler

YALE UNIVERSITY PRESS

NEW HAVEN AND LONDON

This publication is made possible in part by the Barr Ferree Foundation Fund for Publications, Department of Art and Archaeology, Princeton University, and with assistance from the Tina C. Weiner Publication Fund.

yalebooks.com/art

Designed and set in Sabon type by Lindsey Voskowsky

Cover designed by Lindsey Voskowsky

Printed in China by Regent Publishing Services Limited

Library of Congress Control Number: 2019931083

ISBN 978-0-300-24602-5

A catalogue record for this book is available from the British Library.

This paper meets the requirements of ANSI/NISO Z39.48-1992 (Permanence of Paper).

10 9 8 7 6 5 4 3 2 1

Cover illustration: Fabriano, Tamed Horse Watermark (detail of figure 16)

Frontispiece: Andrea Mantegna, “Ludovico Gonzaga Receives a Letter” (detail of figure 53)

Page viii: Cristoforo Roncalli, *Two Studies of a Youth* (detail of figure 69)
Page x: Jacques Daliwe (?) in the *Berlin Sketchbook* (detail of figure 56)

For Theo

CONTENTS

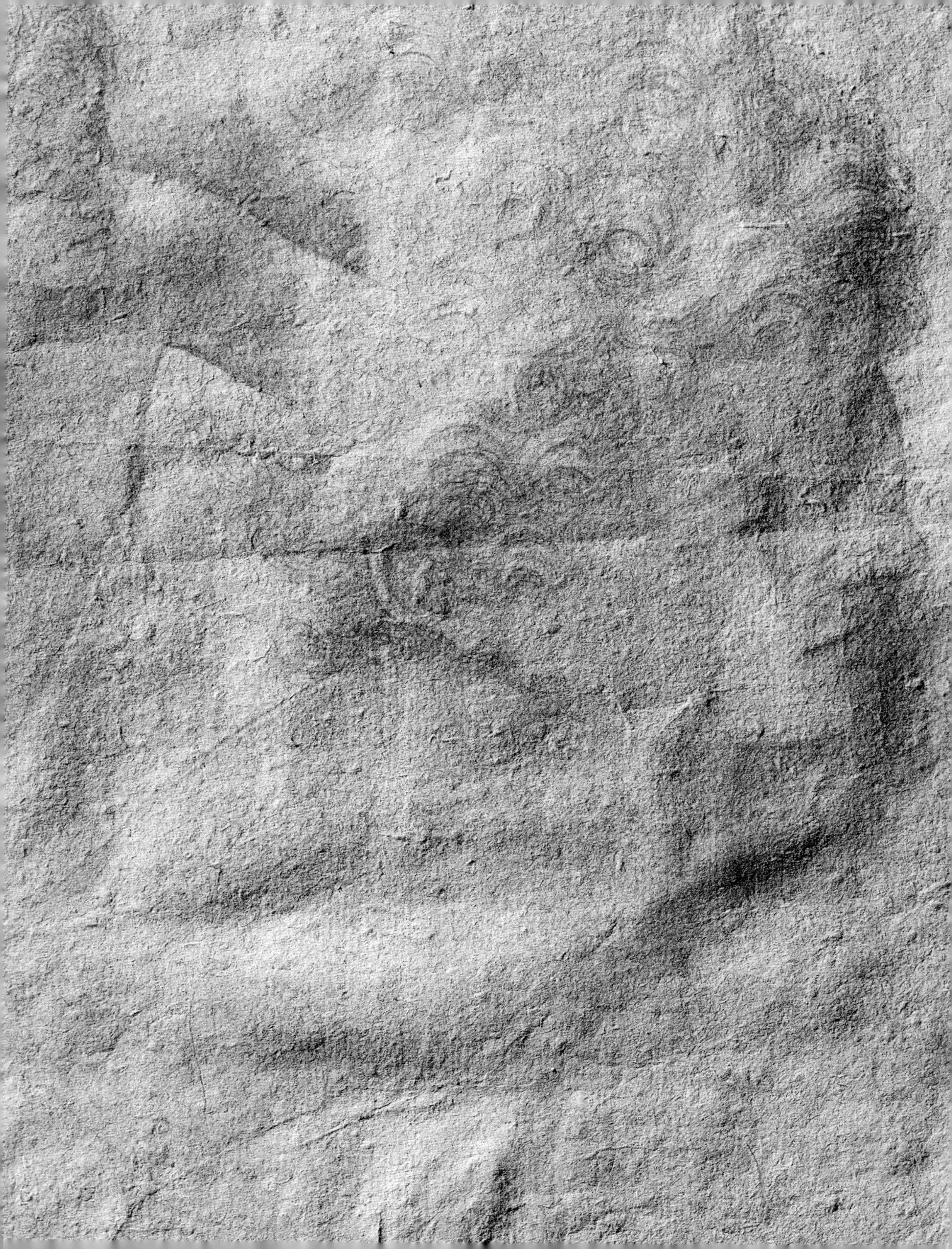

ACKNOWLEDGMENTS

I began this project when I was an NEH Postdoctoral Fellow at the Getty Research Institute, and I thank both the Getty and the National Endowment for the Humanities Foundation for their early support. I completed the manuscript at Yale University as the Andrew W. Mellon Postdoctoral Fellow, and I thank the Mellon Foundation and Tim Barringer. An RSA–Samuel H. Kress Mid-Career Research and Publication Fellowship helped to support the completion of the manuscript. This book was built upon the generosity and expertise of many conservators, curators, and paper specialists who shared their knowledge and collections with me, including Stephanie Buck, Jay Clarke, Barbara Dossi, Margaret Holben Ellis, Theresa Fairbanks-Harris, Oleknka Horbatsch, Franz Kirchweger, Penley Knipe, John Marciari, Christof Metzger, Roy Perkinson, Astrid Reuter, Stephanie Schrader, and Martin Sonnabend. Alexander Nagel and Scott Nethersole read early drafts of "Forgetting Paper's Origins" and offered generous comments. Ittai Weinryb inspired the chapter "Tracking Paper Routes across the Early Modern World." And Katherine Boller was a vital early supporter of the project.

I thank Megan Williams for organizing the stimulating conference "The Politics of Paper in the Early Modern World." Susan Dackerman generously invited me to a stimulating workshop on the extinction of print, where I presented an early version of this project. A lively panel on paper in the artist's workshop at the Renaissance Society of America's annual conference offered important feedback and discussion, and I thank the participants, Shira Brisman, Donald Farnsworth, Mauro Mussolin, and Camilla Pietrabissa. I also thank Alexander Marr for inviting me to present in the "Genius before Romanticism" workshop and for hosting stimulating discussion. Ulrich Pfisterer, Gauvin Alexander Bailey, and the Zentralinstitut für Kunstgeschichte offered an exceptional summer for research and writing. At the Clark Art Institute, Mike Agee and Teresa O'Toole were important in helping to provide technical photography of works on paper. And I thank my colleagues at the Clark: Lauren Cannady, Deb Fehr, Marc Gotlieb, Kathryn Griffith, Olivier Meslay, and Samantha Page, who supported the completion of this project as I began a new job and embarked on motherhood.

Thank you to my parents and my sister for making impossible worlds seem possible. My best reader remains my husband, Nat Silver. And I finished this manuscript as you came into the world, Theo, and "your bald cry/Took its place among the elements." This book is for you.

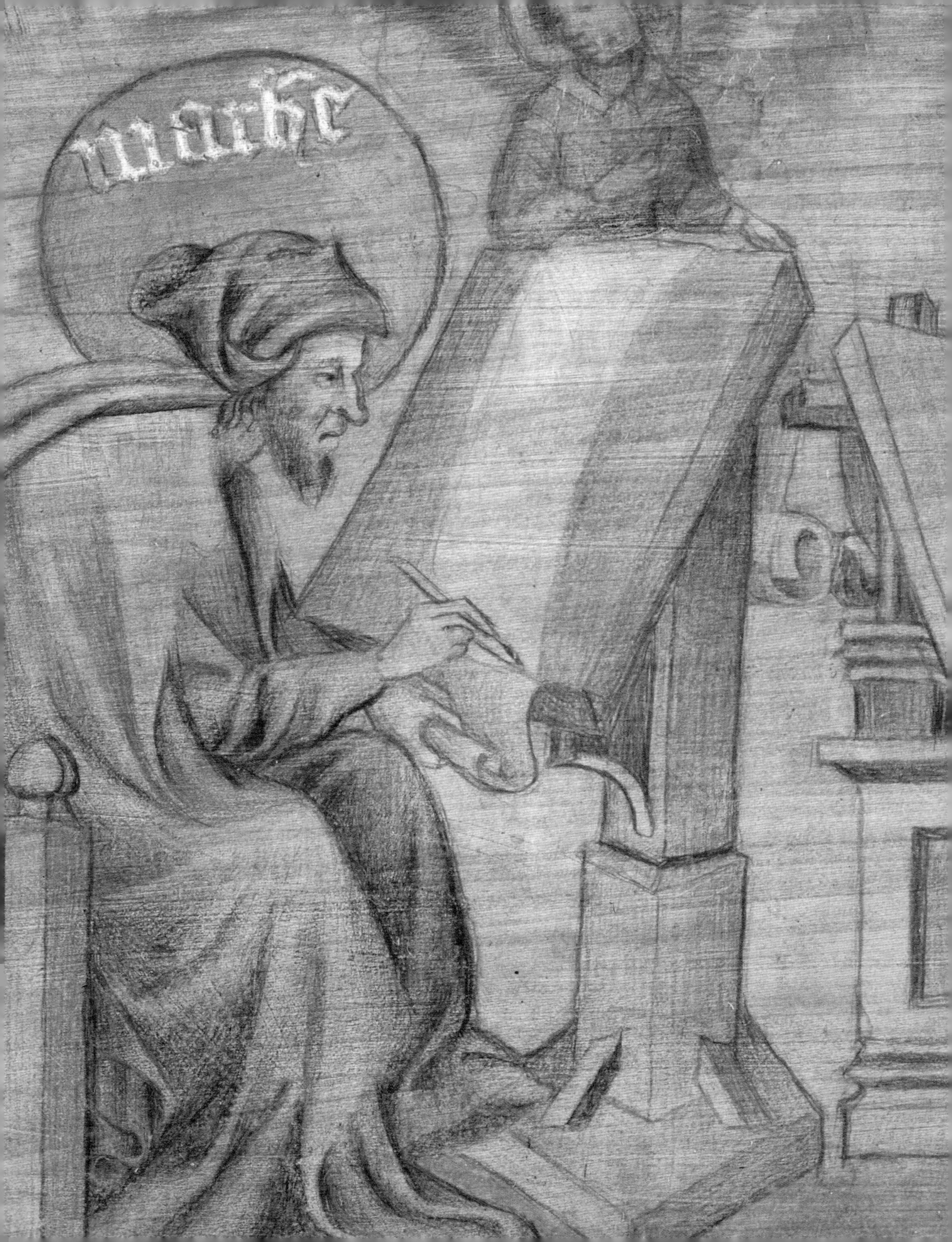
marke

Introduction

A New Kind of Route

"In the year 1525, after Whitsun, in the night between Wednesday and Thursday, I saw this vision in my sleep." Putting mind to paper, Albrecht Dürer (1471–1528) painted his dream, creating a work in which water rises like a geological formation from the horizon as liquid stylolites drip from the sky. He continues: "But when the first torrent to reach the earth was about to strike, it was falling at such speed, with wind and roaring, that in my terror, when I awoke, my whole body was trembling. . . . Then, when I got up in the morning, I painted the above picture, just as I had seen it."[1]

With water and pigment, Dürer materializes the force of his visionary encounter (fig. 1). Yet with his text, Dürer displaced the experience from the atemporal realm of the mystical and set it within the world of sleep and dreams. He specified the year, 1525, and the season, early summer.[2] He noted that it occurred between a Wednesday and a Thursday. And he recorded that he painted it in the morning, after he had awoken. With text and image, Dürer grounded the apocalyptic event within both daily life and the Christian calendar, placing it in relationship to the Holy Spirit descending on the Apostles, and the more banal and diurnal—between a Wednesday and a Thursday. Dürer touches upon the possibility of a vision while he brings it down to human measure. Central to the temporalizing of the dream is the piece of paper on which he recorded the event, marked by tears and rough edges, revealing that it was excised from a journal. Below the transcription is an empty space, standing in for the promise of another drawing, another note, another dream.

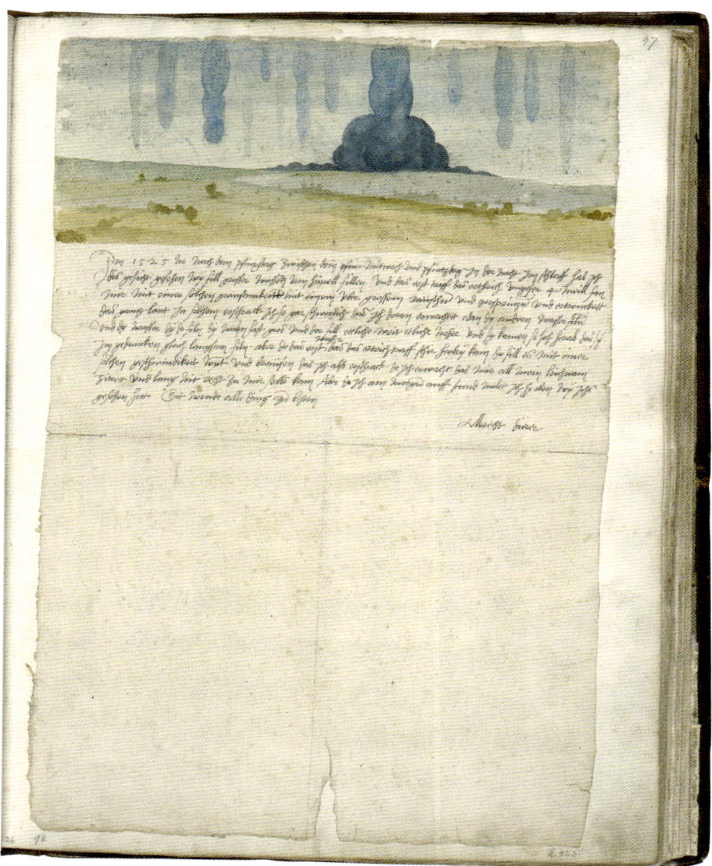

Figures 1 and 2. Albrecht Dürer, *Dream/Vision*, 1525. Watercolor on paper. The Kunsthistorisches Museum, Vienna.

The paper in its entirety is rarely reproduced in Dürer monographs. Instead, the image is often cropped to contain the watercolor and Dürer's script, omitting the accompanying expanse (fig. 2). It is not surprising that scholars do not reproduce the whole sheet, as paper confounds reproduction. Paper is the support for the image, the engraving, the photograph. It is intrinsic to reproductive media, but its functionality prevents it from becoming an image itself. From the late-medieval woodcut to the contemporary photograph, paper is the material on which the image is impressed, captured, or cast, but is itself not the object of representation. The success of other media demands that it appear transparent, denied its own capability to convey a message.

Paper is a three-dimensional object, and it challenges photographic reproduction. Photographing paper in raking light will capture its surface (fig. 3). While photographing paper against transmitted light can apprehend its underlying structure (fig. 4), the surface and structure of paper can never be captured in a single image. This is made clear in the volumes on papermaking by the historian Dard Hunter, in which he always included multiple samples of paper from around the world (fig. 5).[3] Paper can only be conveyed by paper.

The artist who mounted Dürer's vision in an album of his engravings, woodcuts, and drawings considered the entire sheet integral to the work, despite its absence from later photographic reproductions. It is likely that a member of Dürer's workshop, perhaps Hans Döring, assembled the album in the sixteenth century. In compiling Dürer's graphic oeuvre into a bound volume, the editorial artisan trimmed some works, such as *Nemesis,* or the small mythological watercolors (figs. 6, 7), but he kept intact the sheet with Dürer's *Dream*.[4] The presence of the paper therefore also frames the authenticity of the drawing and marks it as a direct trace of Dürer's hand. Works such as *Dream* indicate a shift in the economic and intellectual value of not only drawings but also

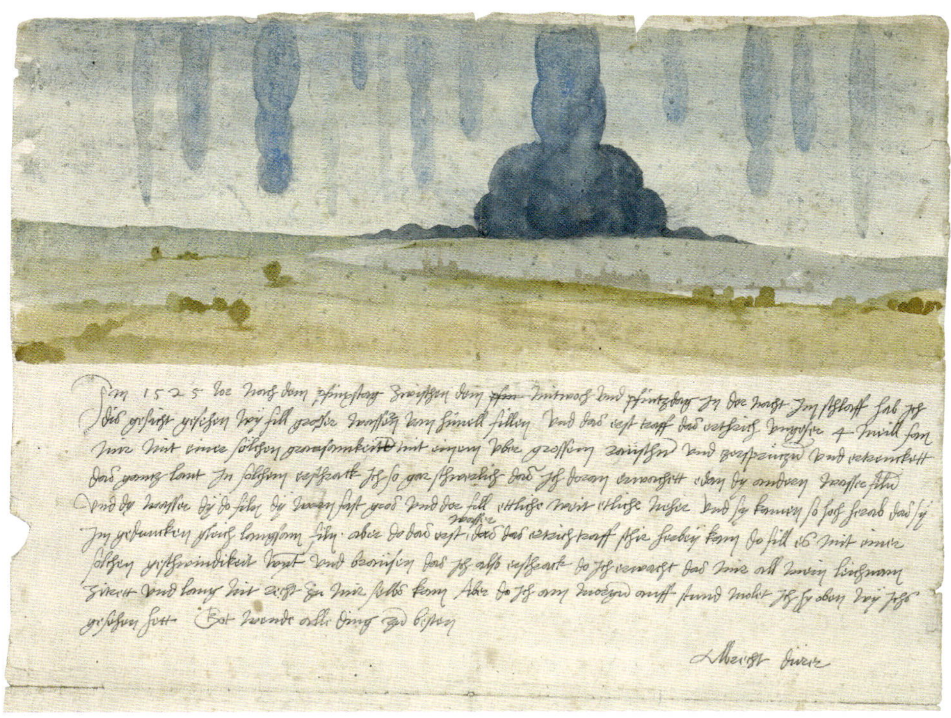

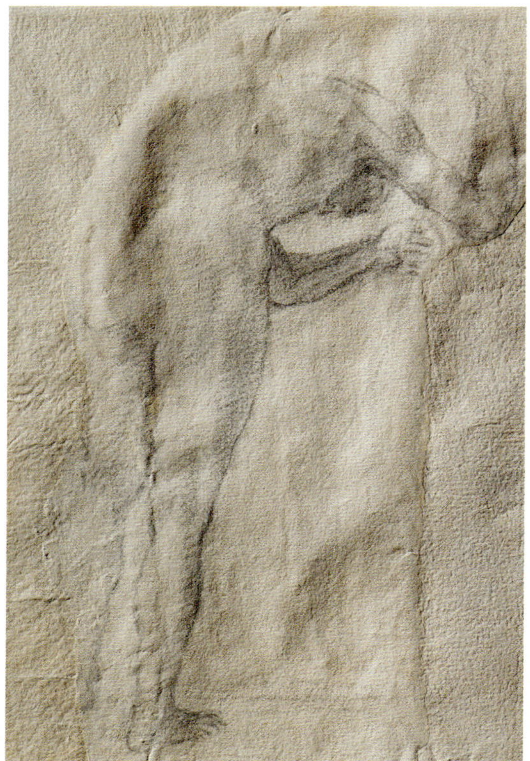

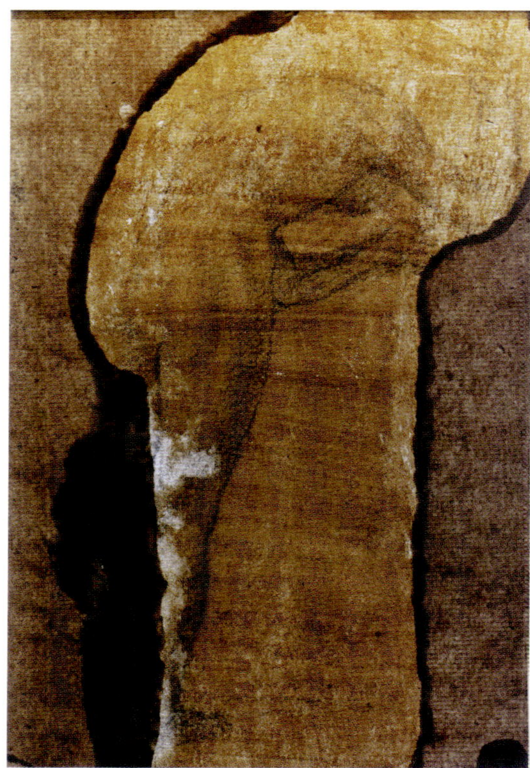

Figures 3 and 4. Timoteo Viti, *Study of a Nude*, 1479–1523. Black chalk with touches of white on paper. Clark Art Institute, Williamstow, Mass. Left: raking light; right: transmitted light.

the artist's more autobiographical notes and sketches. In the sixteenth century, drawing, in particular, became mythologized as Giorgio Vasari (1511–1574) described a Michelangelo (1475–1564) drawing as a relic, suggesting a direct correlation between Michelangelo's body and his work on paper. Like a saint's fingernail or a shard of bone, drawing represented evidence of the draftsman's physical presence.[5] *Dream* occupies an even more ambiguous place as it conveys the workings of imagination and memory in sleep, a liminal activity experienced beyond the limits of the body and the day.

Paper also was ubiquitous enough by the sixteenth century that there was no need for a later artist (or Dürer himself) to employ this precious space. Although it was not to be wasted, paper was neither as costly as parchment or bronze, nor as easy to expunge as parchment or boxwood. The luxurious, unused expanse in Dürer's *Dream* framed it as a fragment, evoking the dreaming

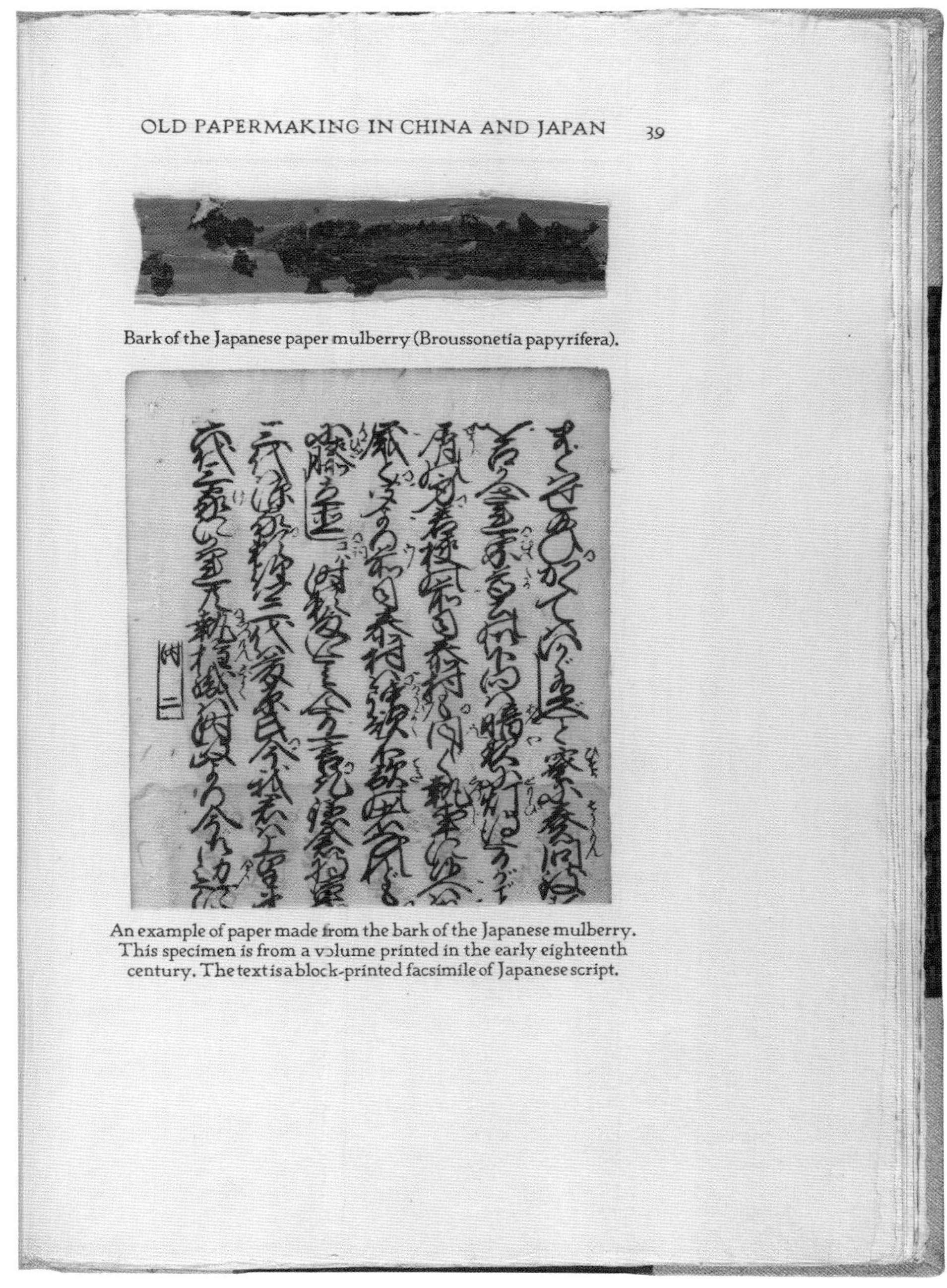
OLD PAPERMAKING IN CHINA AND JAPAN 39

Bark of the Japanese paper mulberry (Broussonetia papyrifera).

An example of paper made from the bark of the Japanese mulberry. This specimen is from a volume printed in the early eighteenth century. The text is a block-printed facsimile of Japanese script.

Figure 5. Dard Hunter, "Samples of Mulberry Bark and paper made from the Japanese Mulberry," in *Old Papermaking in China and Japan*, 1932. Houghton Library, Harvard University, Cambridge, Mass.

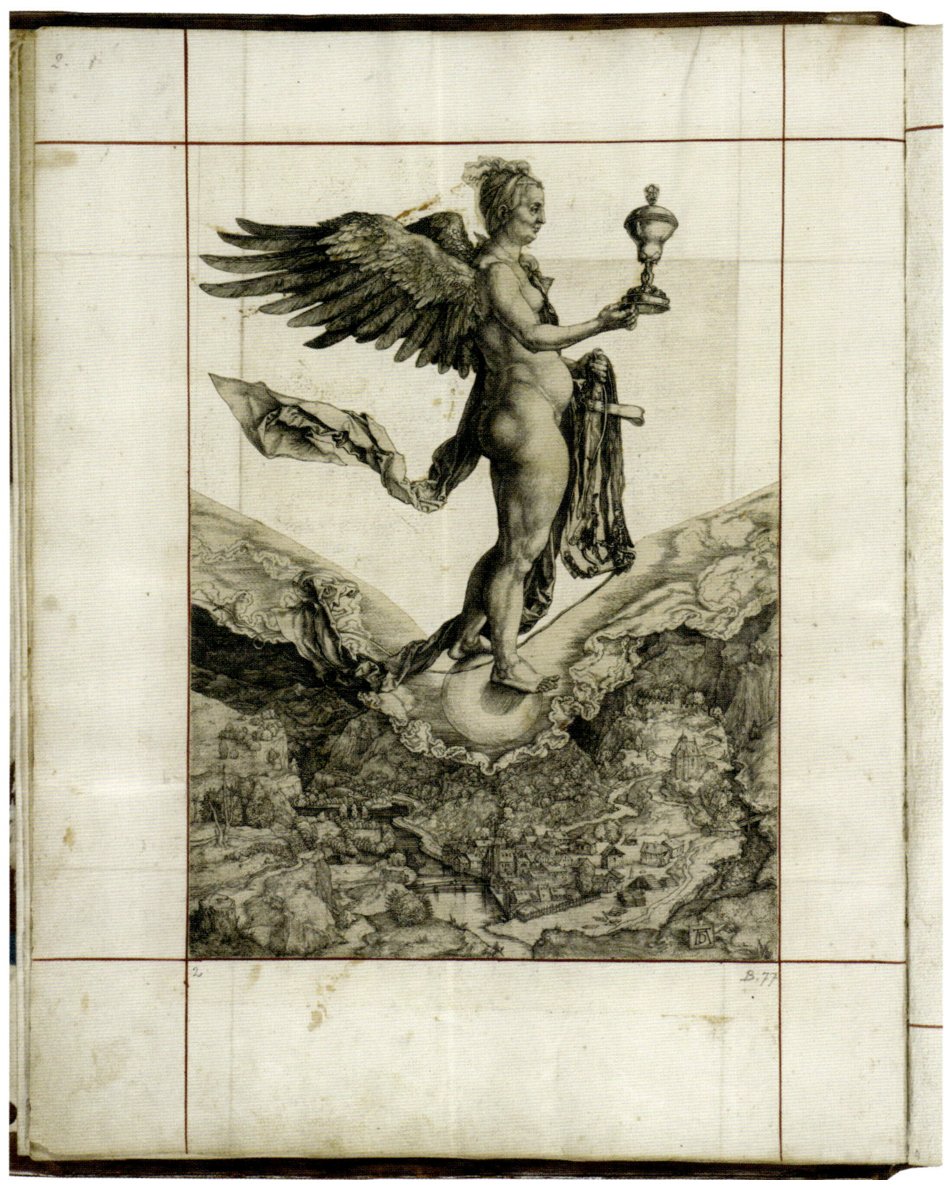

Figure 6. Albrecht Dürer, *Nemesis* (1501), in *Kunstbuch Albrecht dürers von Nürnberg*, ca. 1528–60. Engraving on paper pasted into paper bound book. Kunsthistorisches Museum, Vienna.

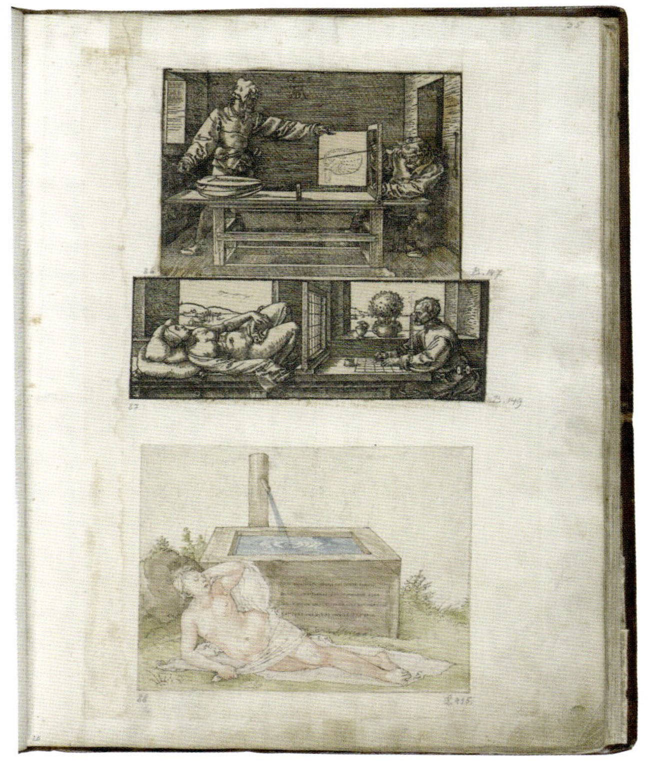

Figure 7. Albrecht Dürer, *Perpective Illustrations* (1525) and after Dürer (?) *Sleeping Venus* (n.d.) in *Kunstbuch Albrecht dürers von Nürnberg*, ca. 1525–60. Woodcut on paper and watercolor on paper. Kunsthistorisches Museum, Vienna.

self as never fully realized. The empty paper suggests that the drawing and text are themselves parts of a larger work from Dürer's self, a labor that by definition remains unfinished, realized only in death. In turn, the void also evokes Dürer's death, reminding the knowledgeable collector that the artist died only a few years after this vision, in 1528. Perhaps, had Dürer lived longer, he would have returned to his journal and drawn another image, or included an addendum. Instead, the paper remains empty.

Although rag paper had been in circulation for at least three hundred years in Europe, Dürer was one of the first artists to articulate its potential as the material of the archive. In my understanding of the archive, I draw on the work of Alexandra Walsham, who points out that early modern archives are not objective sites of collected information, but fluid and dynamic structures through which "actors have exercised and negotiated agency, identity and power."[6] Paper archives were fundamental to artists taking control of their own identities and mythologies, crafting new concepts of practice and authorship. As Michel

Foucault stated on the rise of the author, "Finally, the author is a particular source of expression that, in more or less completed forms, is manifested equally well, and with similar validity, in works, sketches, letters, fragments, and so on."[7] While Foucault talks abstractly about the presence of "sketches, letters, fragments," paper was the physical material on which these were kept. Paper carried the designs, drawings, contracts, letters, and notes that Dürer, and many of his contemporaries, made integral to their legacies.[8] Whereas Dürer purposely saved works on paper, Michelangelo destroyed his. As Vasari relates, Michelangelo burned the drawings that revealed his labor.[9] This story demonstrates a distinct awareness of curating one's paper legacy. The dichotomy between Dürer's collection and Michelangelo's fire expresses the dual qualities of paper as at once strong and enduring, while also flammable and ephemeral.[10] It also points directly to artists' awareness of the fundamental importance of works on paper in shaping their image for posterity.

The paper that transformed the landscape of European art was rag paper, made from the reconstituted refuse of linen and hempen textiles, such as bed linens, sails, rigging, ropes, and clothing.[11] Ragpickers traveled the countryside collecting discarded textiles, which they brought to the paper mill, where they were retted and pulverized with water. The resulting slurry would form a sheet when the suspended fibers were caught in a wire mold. Paper, therefore, was a product of waste and filth.[12] And yet it was also described by the German alchemist Michael Maier (1568–1622) as an alchemical substance, rising like "phoenix out of her own ashes."[13] Paper could be mobile and light, but as the bibliographer Allan Stevenson points out, books in bulk are "weighty things."[14] Paper is at once transient and permanent, mobile and rooted, economical and, as any early modern printer would argue, exceedingly expensive.[15] This is an opposition in paper objects that Jacques Derrida articulates as "the condition of a priceless archive" versus paper as "the throwaway object, the abjection of litter."[16]

Beginning with Lucien Febvre and Henri-Jean Martin's *L'Apparition du livre* (1971; *The Coming of the Book*), historians have attended to the ways in which the production of paper in Europe was a necessary condition for the emergence of print.[17] Yet as Jonathan Bloom demonstrated in his work on Islamic paper, it was a vital commodity throughout the world before the European invention of movable type.[18] Nevertheless, histories of Western paper often are written through the valence of print and therefore present a Eurocentric perspective defined by Johannes Gutenberg. While historians use the term "paper world" to discuss post-Gutenberg Europe, late-medieval Europe already was a paper world, as demonstrated by the works remaining in archives today.[19]

A Maya codex on bark paper survives from 400 to 600 CE.[20] One of the oldest Christian manuscripts on paper from circa 800 is from the Christian community in Damascus.[21] Fifty years later, a scribe in Arabic apologizes for using papyrus and not paper.[22] The oldest book printed on paper, the *Diamond Sutra*, is from the same period (fig. 8). By 970 paper playing cards and currency were circulating in the Song Dynasty.[23] And in 978 a scribe in Syria wrote out the *Arabian Nights* in Arabic on paper.[24] These objects constitute a minute sliver of the paper traces left around the world before the first paper mills were established in thirteenth-century Italy.

The Origins of Paper

Papermaking was invented in China and traveled via Samarkand into the Arab world (fig. 9).[25] By the eighth century, it was a prominent writing support in the Abbasid Empire, where it also underwent a significant technological shift.[26] Although there is evidence of early paper made from rags, paper in China was predominantly composed of plant material, such as mulberry.[27] The tradition of making paper on a large scale from the reconstituted fibers of linen and hempen rags took shape under Arab papermakers in Samarkand. Paper did not enter Europe until about the eleventh century, via North Africa into Spain.[28] When the Muslims controlled Spain, they established a papermaking industry.[29] With the rise of the notary in Europe, there developed a need for a more economical support than parchment, and paper began to travel from Spain and into Italy, France, and Germany. At the end of the thirteenth century, records establish the first paper mills in northern Italy.

The rise in administrative governance and bureaucracy was crucial to paper's success across cultures.[30] Paper became dominant as a support under the Abbasids because it was cheaper and more abundant than papyrus and parchment, and therefore able to meet the increasing organizational needs of the caliphate.[31] In turn, European notaries adopted paper for their record-keeping, instigated by the revival of Roman law. Paper became integral as a tool of administration not only because it was more economical than papyrus and parchment, but also because it was more difficult to forge than a parchment document. Animal skin could be scraped down and erased. By contrast, paper resists erasure, and tears before the writing disappears. The fundamental role of destruction in parchment culture—and in turn the ways in which its palimpsestic qualities impacted the survival of texts—is illustrated by one of the remains of Pliny's writings from late antiquity.

Pliny died in the eruption of Mount Vesuvius in 79 CE, but his earliest texts date from the fifth century. Although Pliny likely wrote his manuscripts

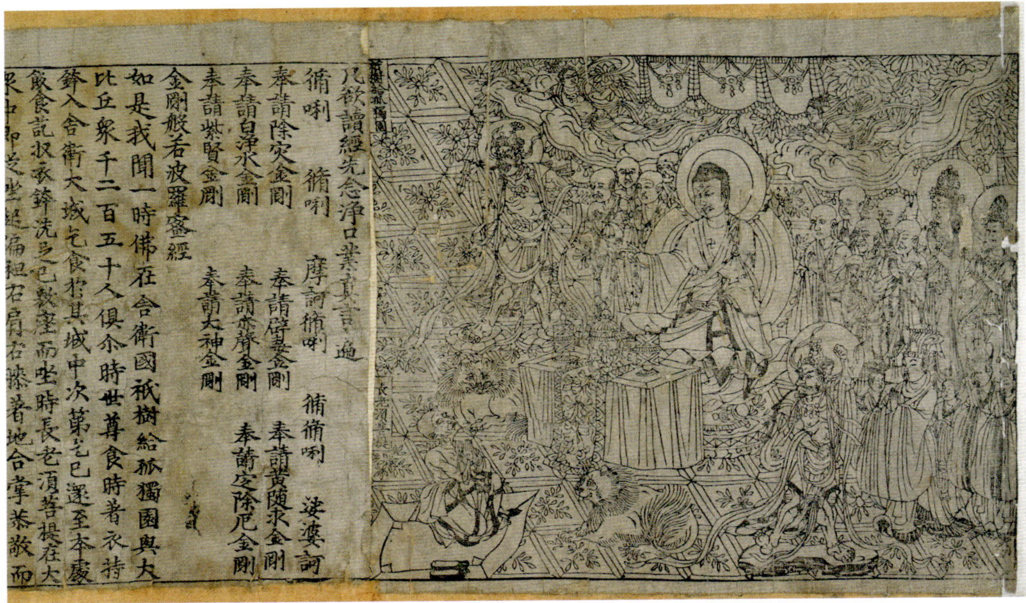

Figure 8. Diamond Sutra, 868 CE. Woodcut on paper, 11 in. × 16 ft. (27 cm × 5 m). British Library, London.

on papyrus scrolls, the surviving versions are palimpsests found in parchment codices. One of the foremost editions is the fifth-century *Codex Moneus,* discovered in a Carinthian monastery library (fig. 10). It contains chapters 11 to 15 of the *Naturalis historia* (*Natural History*), buried under another script by a later hand. The durability and thickness of the parchment allowed the seventh-century scribe to scrape away Pliny's work and to copy a new text, St. Jerome's *Commentarius in Ecclesiasten* (*Commentary on Ecclesiastes*).[32] Nevertheless, the imprint of the original may be discerned beneath Jerome's *Commentarius.* Pliny's textual ruins are not buried in the earth as lapidary inscriptions, but they must be uncovered.

Paper, by contrast, preserved in a way that parchment did not. Parchment was thinned and reused. It was also expensive and not favorable to experimentation. Paper lends itself to keeping, filing, forgetting, and finding.[33] Although the rise of European paper is invariably interpreted in relationship to the history of print, paper was transformative outside of typography and the engraver's plate. The importance of paper to artistic practice must be seen not only in the context of print, but also the rise of the late-medieval notary, the emergence of

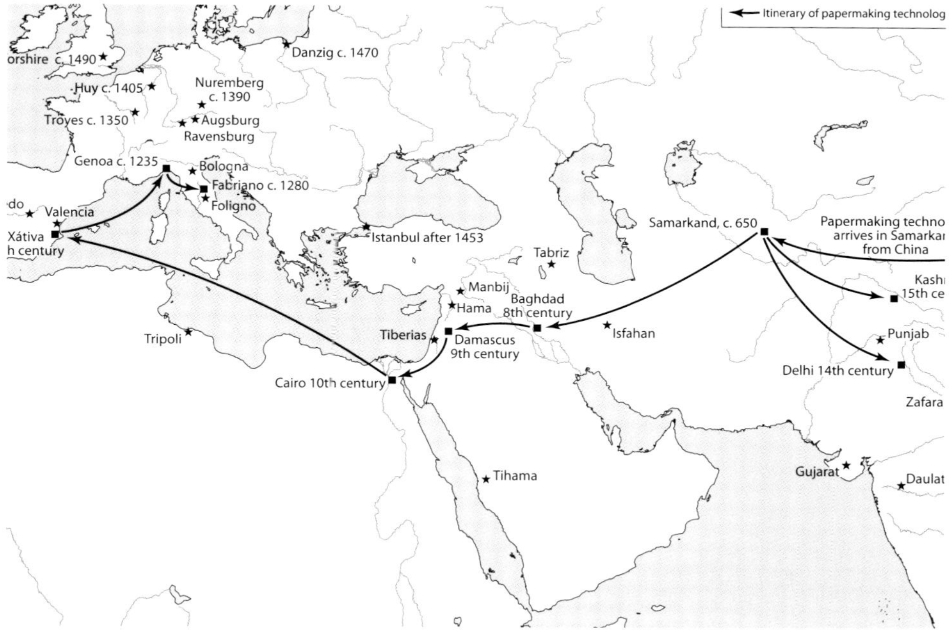

Figure 9. Itinerary of rag papermaking technology from ca. 650–1500. Design by author, cartography by Bill Nelson.

the first major European archives, and the first European paper governments, from the Holy Roman Empire of Maximilian I to the bureaucratic structure of New Spain, governed by Philip II. As the media theorist Harold Innis argued, paper was pivotal to events that define the early modern world, from the revival of Roman law to the emergence of the nation-state.[34] Paper's deployment from China to New Spain shares a common feature of centralizing an ever-expanding state bureaucracy, a tension between expansive circulation and the particular site of the state archive. Artists who feature in this book reveal an acute awareness of not only the diffusive multiplicity of paper works, but also of the singular, irreplaceable, and site-specific logic of the handwritten document. Yet just as paper captured a new relationship to space, it also arrested a new temporality.

Paper Time

Two of the earliest examples of Italian drawings on paper, once attributed to Simone Martini (1284–1344), illustrate the shifts that paper allowed in artistic practice.[35] The recto depicts Christ and the woman of Samaria at the well,

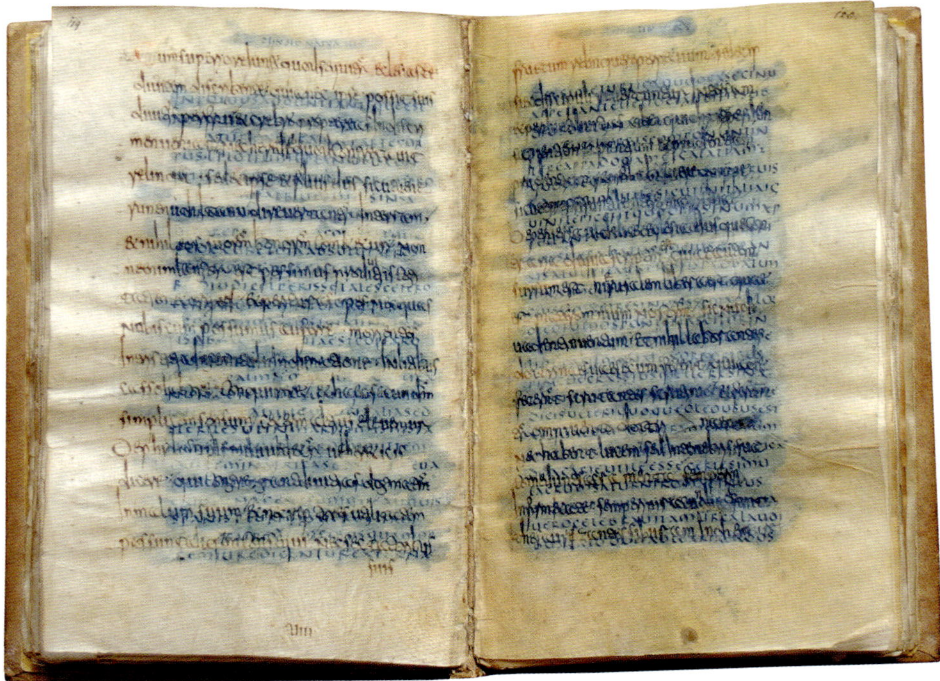

Figure 10. Jerome, *Commentarius in Ecclesiasten*. Palimpsested: Pliny, *National History, XI–XV*, in *The Codex Moneus* (Archiv: Cod. 3/1), 5th century/8th century. Ink on parchment. Library of the Benedictine Abbey of Saint Paul's in Sankt Paul im Lavanttal.

and the verso represents Christ healing the blind man (figs. 11, 12). Although the drawings cannot be connected to a specific work, they are attributed to a Sienese artist in the fourteenth century due to their similarities to Duccio's *Maestà*. Both images were first drafted with black chalk and then later reinforced with pen, brown ink, and wash, and heightened with lead white and red chalk. The architectural scene and landscape are delineated in broad, quick strokes, while the primary attention is accorded to the arrangement of the figures, their gestures, and their bodily expression. The tree in the background is circumscribed with rapid parabolic lines. Similarly, the architectural structures display a perfunctory interest in perspective. The drawings, in this cursory fashion, share a logic with the registry, keeping a record of the compositional act in its briefest form.

On the verso, the paper surface arrests the indecision of the draftsman as he shifts Christ's hand. First, the artist sketched Christ blessing the blind man,

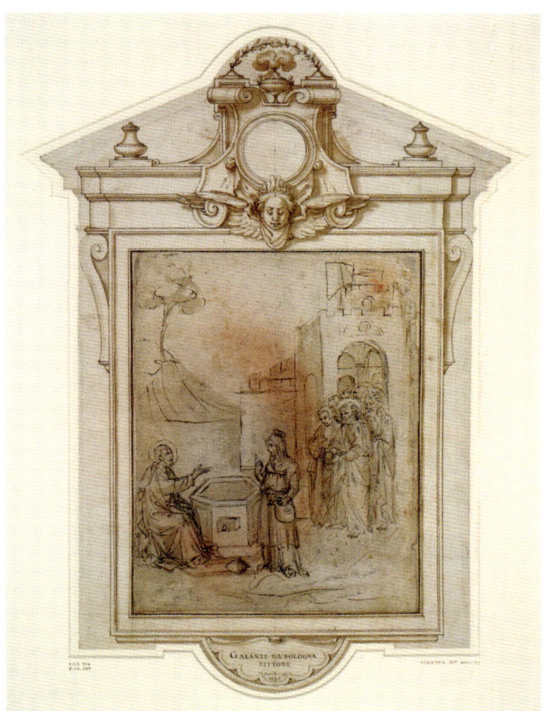

Figure 11. Anonymous (Sienese?), *Recto: Christ and the Woman of Samaria,* ca. 1390–1410. Pen and brown ink with gray-brown wash, heightened with white over black chalk on paper rubbed with red chalk, mounted in a later period either within or in imitation of Giorgio Vasari's *Libro de'Disegni.* The British Museum, London.

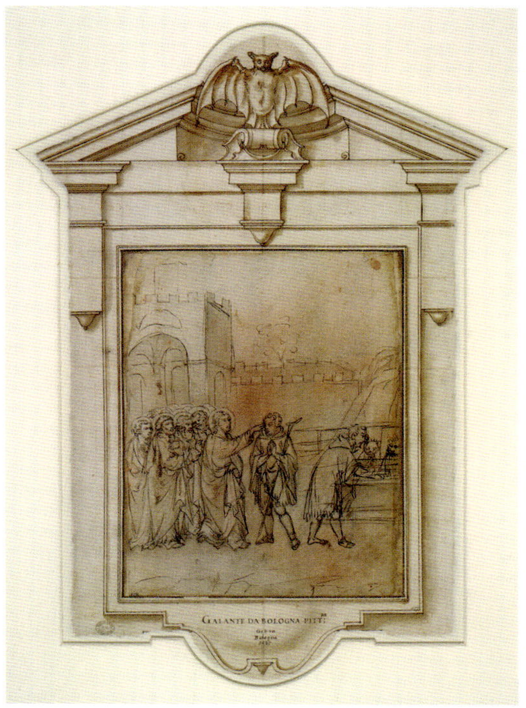

Figure 12. Anonymous (Sienese?), *Verso: Christ Healing the Blind Man,* ca. 1390–1410. Pen and brown ink with gray-brown wash, heightened with white over black chalk on paper rubbed with red chalk, mounted in a later period either within or in imitation of Giorgio Vasari's *Libro de'Disegni.* The British Museum, London.

a gesture that the artist changed to Christ touching the man's closed eyelids.[36] Paper offered artists the opportunity to reconsider the position of a hand—the transference from blessing to touching—with greater freedom. Paper apprehended the transitions as the draftsman worked, seizing thought.[37] Preparatory drawing was a practice entrenched in the workshop before the arrival of paper. But these sketches were drafted on erasable surfaces: wax tablets, parchment, or directly on wood panels. The novelty of paper was its ability to document the movement of the draftsman as he attained (or failed to realize) the desired form.

Art historians have ascribed a new awareness of temporality, as seen in Dürer's demarcation of dates, to the invention of print, and its ability to set a rapid pace by which images were both produced and circulated. Yet in her work on late-medieval legal documents, Cornelia Vismann demonstrated a different consciousness of time, one that did not arise with print but with paper. She notes that when the thirteenth-century court of Frederick II in Sicily adopted the daily paper register, it allowed the development of blank spaces in paperwork.[38] Scribes began using a new page at the beginning of each month, which created a spatialization of time, as discrete dates could be easily retrieved by the user of the registry. This new paperwork "removed power from the realm of eternity and subjected it to time." Through displaying the course of *acta*, the registry "made visible the continual flow of time."[39] Vismann's analysis demonstrates the role of paper before print as central to making legible a linear progression of actions, and contextualizes an emergent temporal rhythm made visible by paper in the practice of drawing.[40] This approach to paper time also informs the logic of Dürer's *Dream*, as the blank space secures the vision within the specific interval described by Dürer.

This book will highlight a series of junctures in the early history of European paper, demonstrating paper's trajectory from its initial thirteenth-century arrival in Europe as an import marked with traces of the Judeo-Arabic world to its export in the sixteenth century to New Spain. By the time the Spanish were utilizing paper for the administrative governance in the New World, the support was seen as an embodiment of European ingenuity, and a tabula rasa for invention, its previous Judeo-Arabic history forgotten. In order to lay the groundwork for this paper history, the first chapter introduces a transcultural history of paper. In the seventeenth and eighteenth centuries, a literature in print about papermaking developed in both China and Europe. This created a legible iconography of the technology, which allowed for comparisons between Asia and Europe, a comparative history that in turn erased other prominent paper economies, namely the Islamic empires and Mesoamerica. As this chapter maintains, paper technologies formed outside of print cultures were nearly obliterated by the printed matrix and by the translation of manuscript culture into print. In this movement across media, there was little account for the material histories of the book lost in the processes of conversion.

The second chapter maps the early itineraries of paper into Europe as a commodity oriented toward the Holy Land. Paper was associated with Islam, Judaism, and a history of the book that was distinctly not Christian. Paper was Christianized and commodified gradually through both the invention of the watermark and the printing press. Although this early history of paper in Europe was quickly forgotten, artists such as Simone Martini and Andrea Mantegna (1430/31–1506) actively engaged with the non-Western origins of paper at the same moment that paper was attaining its status as a commodity signaling power within European diplomatic circles.

The third chapter considers the introduction of paper to the illuminator's workshop, and its coexistence with the supports of parchment and boxwood. Within the context of the court, paper revolutionized the authority of the lyricists and clerks. With paper, these writers and bureaucrats laid claim to a new sense of authorship through the physical formation of their literary canons, an emergence of the *auctor* that would come to influence artists. It is within the survival of late-medieval model books on boxwood that the first notion of the artist's physical legacy as a paper archive emerges.

The fourth chapter examines Albrecht Dürer, the artist who first grasped the importance of paper as a material with which to build not only a history of draftsmanship and style, but also to define his own reception. Although prints are vital to Dürer, his drawings reveal an artist shaped as much by the logic of paper as print. Unlike his engravings and woodcuts, where the matrices conceivably could be struck anytime and anywhere after his death, his drawings carried the site of their making in a way that was particular to the paper support, a geographic specificity that Dürer exploited in his travels between Nuremberg and Venice.

Dürer's understanding of the paper surface and its ability to embody both place and mobility traveled outside of Europe and into New Spain. The final chapter analyzes the coexistence of European and indigenous paper in New Spain, demonstrating that the support of legal documents, letters, maps, engravings, and drawings was not a transparent bearer of knowledge. It embodied divergent forms of knowing, which in turn impacted how the world was drawn. With the rise of colonialism and shifting notions of property and self, paper became mythologized as an empty plot primed for inscription. The blank space was no longer the preserve of the notary's registry, or as seen in Dürer's *Dream,* the means to arrest time. Instead it became an idealized space in which both to form one's intellect and to define borders, imagined and real.

Early modern artists were cognizant of the geographic variety of paper, and the ways in which it grounded specific ideas about memory and imagination that existed in relationship to other supports, such as bronze, parchment, wax, and wood. Yet the meaning of paper changed in the seventeenth century. Paper became aspatial and atemporal. It was either inscribed or empty (a tabula rasa). It was either European or other. The empty sheet became a placeless place, a virtual space, a transparent bearer of knowledge.

This is a history composed of documents and drawings, legal treatises and philosophical tracts. These paper works coexisted and informed one another as paper marked new routes both within Europe and across the Pacific and Atlantic oceans. Papermaking could symbolize the washing of the soul, the cleansing of bodily waste embodied in the filthy, discarded rags. In turn, this material—both alchemical and mundane—conveyed the knowledge of the heavens and the stars, Earth's journey around the sun, the anatomical and the diseased, the microcosm and the macrocosm, and the distant terrains unknown to Ptolemy. Paper became the support to describe, map, and know the early modern world, charting a route, always accumulating, seemingly infinitely expanding, only to be filed away, archived, and kept for posterity, recording a new knowledge not defined by divinity but by the world itself.

CHAPTER 1

Tracking Paper Routes across the Early Modern World

Before Europe began producing its own paper, three papermaking economies dominated the world from about 800 to 1250 CE: Asia, the Middle East, and Mesoamerica. Paper was central to the trade, economy, and artistic cultures of China, Japan, and Korea. It was the main component of bookmaking and administration in the Islamic world from Baghdad to Córdoba. And it was a material revered by both the Maya and the Aztec. The papermaking technologies across these distinct areas and cultures differed significantly, but it was Europeans' encounters with all markets that shaped the emergent meaning of rag paper in the early modern world.

Paper is made from most plants. Cellulose, a key structural component of botanica, releases glucose when broken down, and those molecules bond to the cellulose, forming a new tensile surface. In Asia, paper was made with barks and vegetation from bamboo to seaweed. In Mesoamerica, it was created from the many species of fig trees and agave plants, among other indigenous botanica. And in Central Asia, papermakers developed the art of manufacturing paper from linen and hempen rags. This latter technology depends upon textiles made from plants, such as flax woven into linen. When the fabrics break down with physical pressure, water, and sometimes chemical agents such as lime, they release the same cellulose and glucose molecules as the raw plant materials. It was this rag-paper technology that spread throughout the Islamic world, and into Europe.[1]

Paper *Technê*

One of the first comprehensive illustrated treatises on European rag paper, *Art de faire le papier* (1761; *The Art of Making Paper*) by Jérôme de La Lande (1732–1807), contextualizes the production of paper within the proto-industrial mill economy of early modern Europe, while also offering a brief comparative history of European paper with its Asian counterparts.[2] La Lande was not a papermaker but an astronomer. He dedicated his scholarship to the heavens and the sister-art of navigation, developing a rigor in observation that he applied to his analysis of papermaking. The treatise was part of a larger project, known as *Description des arts et métiers* (1761–88; *Description of Mechanical and Industrial Arts*), which unfolded under the aegis of the Académie royale des sciences. The *Description* involved members of the academy observing and publishing on the industries of the French economy, covering a variety of topics, including the production of metals, scientific instruments, ships, textiles, porcelain, pottery, and sugar. Although the *Description* is not as renowned, many of the engraved plates were the basis for Denis Diderot's illustrations in the *Encyclopédie* (1751–66).[3]

La Lande detailed both the actions of the papermakers and the technical design of the mill. He began his illustrations with the women who classified the rags into varying qualities: fine (*les fins*), medium (*les moyens*), and coarse (*les grossiers*).[4] The narrative focuses, however, on the structure of the paper mill, and the technology employed to channel running water to transform soiled linens into a new material (fig. 13). In his portrayal of the engineering, La Lande illustrates papermaking as a technology that embodied man's ability to transform the raw resources of the natural world into an artificial product that synthesized nature.

La Lande's attention to the architecture of the mill demonstrates the importance of waterpower in driving local economies. Mills were central to economical changes in late-medieval Europe, as the historian Marc Bloch maintained, arguing that they produced a proto-Industrial Revolution.[5] Paper is essential to this story, for the advancements in its industry contributed to Europe both cornering the paper market and changing the structure of the sheet.[6] The rag-paper trade in Europe transformed the global production of paper into a European symbol of ingenuity, technology, and mastery over nature.

This command over the natural world is the significant attribute of paper, according to Francis Bacon in his *Novum Organum* (1620; *New Organon*), in which he expands upon the particularity of paper in relationship to other man-made things, including silk, wool, linen, glass, earthenware, enamel, and porcelain. In contrast to these other products, paper is strong and durable because, like skin, it will hold together even while being cut and torn: "Paper however is

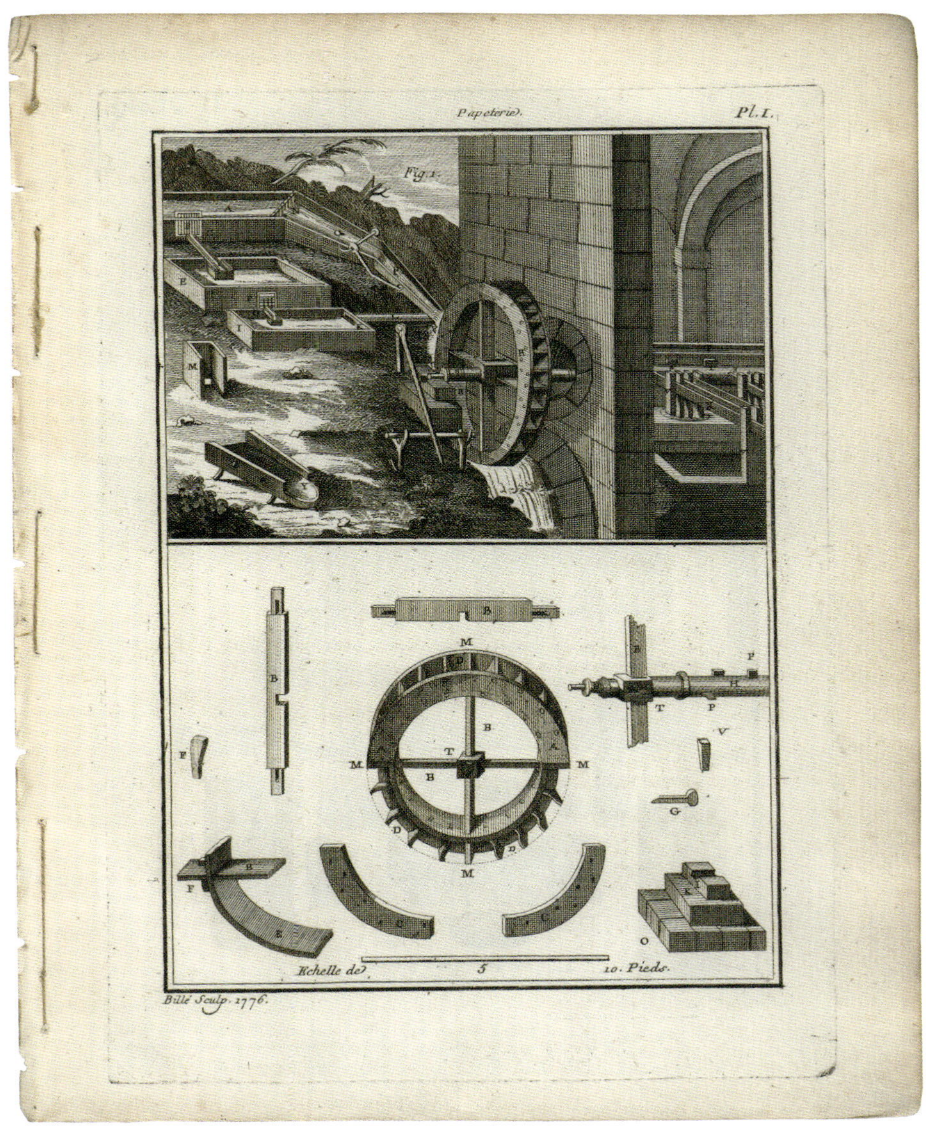

Figure 13. Jérôme de La Lande, "Mill on the Grand River Auvergne" in *Art de faire le papier,* 1761. Engraving on paper. The plates are signed: "Billé sculp. 1776." Houghton Library, Harvard University, Cambridge, Mass.

a tenacious substance which can be cut and torn; so that it imitates and almost rivals the skin or membrane of an animal, or the foliage of a vegetable, and such natural products. It is not fragile like glass nor woven like cloth; it certainly has fibres, but not distinct threads, altogether like natural materials. And so it is found to have hardly any similarity with other artificial materials but to be absolutely unique."[7] Bacon marvels at paper's structure as a continuous surface. He suggests that paper is a "better kind" of artificial material because it not only imitates nature, but also could "masterfully rule her and change her completely."[8] And it is this ability to control both nature and the physical landscape that dominates La Lande's treatise, a jurisdiction reflected in the continuous and regulated surface of paper.

La Lande takes his reader into the paper mill, where the water powers the hammers that stamp and beat the rags (fig. 14). He explicates the process of breaking down the textiles, which occurs over a series of stages. In the first phase, the textiles are shredded, a process that can last six to twelve hours, expunging all traces of the linen and hempen rags. The hammers for this action are wooden wedges with sharp iron nails. In the second juncture, the rags enter into the beater troughs, where they are processed for twelve to twenty-four hours with flat-headed nail hammers. And in the final step, the slurry is "brushed" with hammers only composed of wood, liquefying the pulp. Once the slurry reaches its desired consistency, the process of making a sheet begins.

One of the first modifications to rag-paper production introduced in northern Italy was the alteration of the stamper heads that beat the textile fibers into a pulp. Fourteenth-century papers are distinguished by a lumpen surface with twisted bits of hempen or linen fibers (fig. 15). The uneven texture prevents a clear reading of the laid and chain lines from the paper mold when the sheet is held up to light. Yet throughout the fourteenth century, the pulp became finer as the stamper heads were altered, and the nails in the trip-hammers were modified to increase productivity.[9] In the fourteenth century, rogue fibers and yarns became increasingly rare, so that if a sheet of paper is held up to the light, it is a transparent carrier of both its mold and watermark (fig. 16).[10] The increasing liquidity of the paper pulp allowed the finished sheet to become a bearer of its own technology, conveyed by the watermark and the gridded mold. By the seventeenth century, rag paper reached a pinnacle, as the fine wires were densely packed together into laid lines evenly cut by the clear demarcation of the chains. In turn, the watermark attained a new complexity indebted to both the fineness of the pulp and developments in wiredrawing technology (fig. 17).[11]

As in the modification of the stamper heads, Europeans also improved the art of wiredrawing in the fourteenth century, making thin, strong, and consis-

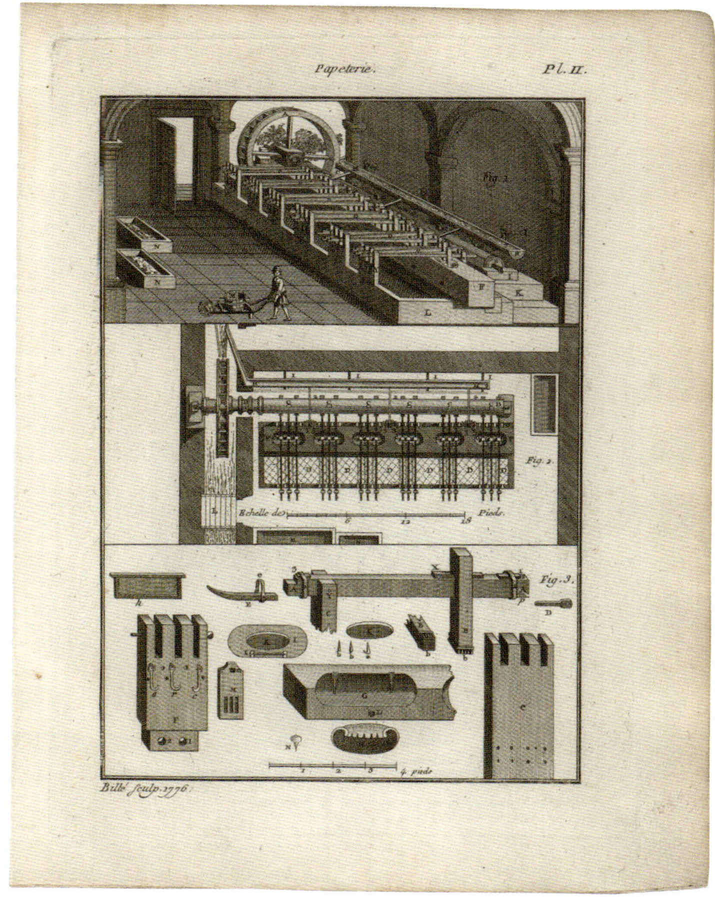

Figure 14. Jérôme de La Lande, "Papermill Stampers" in *Art de faire le papier,* 1761. Engraving on paper. The plates are signed: "Billé sculp. 1776." Houghton Library, Harvard University, Cambridge, Mass.

tent wires. It was this innovation that led to the formation of paper molds with finer chain lines and laid lines. The consistency of the wire paper mold supported a strong and even paper surface, impressed with the mold's grid. It is not a coincidence that water-powered wiredrawing technology was developed in Nuremberg at the same time that Ulman Stromer established his famous paper mill in 1390. In one of his earliest watercolors, Dürer depicted Nuremberg's emergent mill sector and the relationship between technology and landscape (fig. 18), marking the drawing with the word "*Drahtziehmühle*" (wiredrawing mill).

With this work on paper, Dürer engaged with two forms of waterpower: the materials for the watercolor and the depiction of a water-powered mill. Yet Dürer attended not only to the mills but also to the landscape in which

Figure 15. Fabriano, Tong Watermark, 1333. Ernst Kirchner Collection, Deutsches Museum, Munich.

they were situated, rendering with impressionistic strokes the blue mountain range while picturing the local buildings with linear precision, documenting the integral connection between industry and environment. His attention to the water-mill quarter reflects not only an emergent interest in landscape, but also a concern with how locale influenced production, drawing a correlation between environment and commodity. This integral relationship between local land and production became embodied in the symbol of the wire watermark, which would carry the mark of the water mill's quality—determined by its location and local water supply—across Europe and into distant worlds. In acknowledgment of this innovation, La Lande dedicated a single image to the paper mold (fig. 19), documenting its construction with a wooden frame, a metal sieve, and a movable frame called the deckle.

Although the mold was the necessary tool, the physical action of the vatman, who dipped the mold into the paper slurry, was equally important (fig. 20). Catching the fibers in order to form a piece of paper demanded two different movements. First, the vatman "shakes" (La Lande uses the term "*promener*") the mold right to left, so as to spread the fibers. Then the vatman must "shut" (*serrer*), a back-and-forth movement, to bind the fibers. As La Lande remarks: "Immediately this stuff, so fluid, which seems no more than slightly cloudy water, knits together, its small particles bind and unite with one another, and without these two movements some would fall back into the vat through the mould."[12] Once the sheet set, the vatman removed the deckle and passed the mold to the coucher. The vatman would then apply the deckle to a second mold and continue shaking and shutting, while the coucher pressed the sheet onto

Figure 16. Fabriano, Tamed Horse Watermark, 1347. Ernst Kirchner Collection, Deutsches Museum, Munich.

Figure 17. Amsterdam City Coat of Arms, 1680. Ernst Kirchner Collection, Deutsches Museum, Munich.

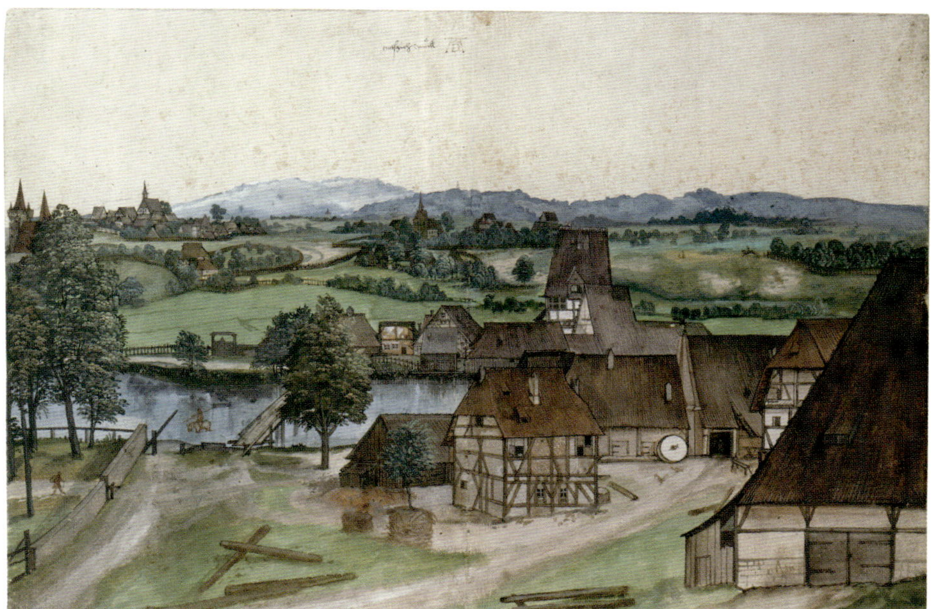

Figure 18. Albrecht Dürer, *The Wire-Drawing Mill*, ca. 1494. Watercolor and gouache on paper, 11 7/16 × 16 3/4 in. (29 × 42.6 cm). Kupferstichkabinett, Berlin.

a piece of felt to absorb the water before dying the sheets in a loft. If the mill produced fine writing paper, it would be sized with a gelatin substance made from animal skin that would make the surface both absorbent (but not too permeable) to the liquid ink and resistant to the quill's sharp edge. The paper was then burnished and packaged for shipment.

In La Lande's *Art de faire le papier*, paper is both a product of human productivity and a commodity formed by technological innovation. He illustrates papermakers at work in well-lit interior spaces, in conjunction with isolated renderings of the mill's machinery. In deploying these two types of engravings, La Lande's treatise enacts the same logic explicated by Roland Barthes for Diderot's *Encyclopédie*. Barthes points out that Diderot's illustrations are in two registers. One is dedicated to the scenes that incorporate machines and industry into daily life, as may be seen in La Lande's opening illustration of women sorting textiles at the mill. The other depicts the machines in detail, and devoid of narrative.[13]

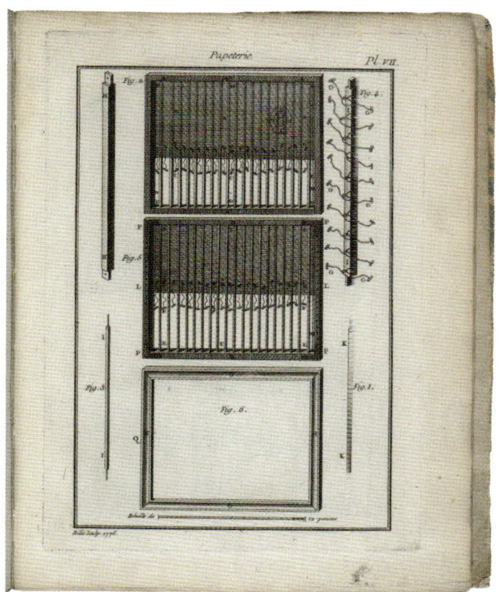

Figure 19. Jérôme de La Lande, "Paper Mould" in *Art de faire le papier*, 1761. Engraving on paper. The plates are signed: Billé sculp. 1776. Houghton Library, Harvard University, Cambridge, Mass.

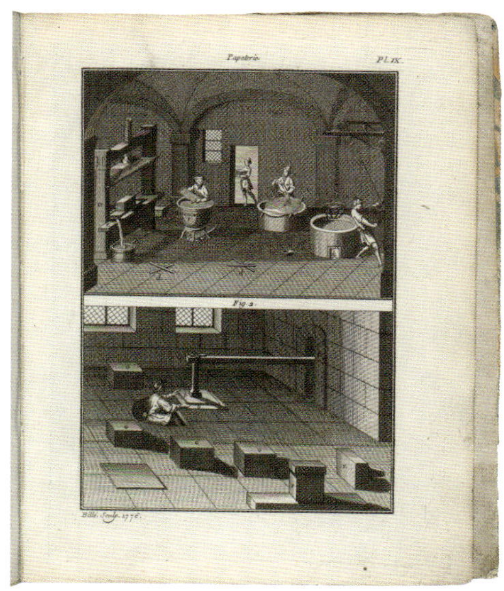

Figure 20. Jérôme de La Lande, "The Vatman and the Coucher" in *Art de faire le papier*, 1761. Engraving on paper. The plates are signed: Billé sculp. 1776. Houghton Library, Harvard University, Cambridge.

Yet there is a third register, not enumerated by Barthes, present in La Lande's papermaking illustrations. In images of shipbuilding or porcelain, the final commodity—such as the galleon or the vase—cannot be present in the final engraving. Instead the objects exist between the representational limits of the engraver's burin and the reader's experience. By contrast, the final product of paper is integral to the publication of the treatise on papermaking. The action of reading the treatise incorporates the economy of print and paper into daily life. Unlike metallurgy, or even the internal organ system of the human body, the paper is visible, touchable, and integrated into the telling of its own story. And yet the reader can also separate the page being turned from the material under explication.

For La Lande's account is not about paper, but *papermaking:* the technology, the mill, and the channeling of water to create. The paper is a by-product of these other forces. The illustrations do not present the object of paper, but the iconographic vocabulary of papermaking. This distinction exists at the heart of a European history of rag paper, long considered a symbol of the transformative properties of water and wood to convert soiled linens into a clean, white surface that, when held up to the light, turned translucent to reveal the structure of wire, grid, and watermark.

Nevertheless, Europe also realized the limits of its own technology as rag shortages disrupted the industry in the seventeenth and eighteenth centuries.[14] Although Europeans noted that many other cultures around the world created paper from the bark of trees and other raw plant material, wood-pulp paper was not invented in Europe until the nineteenth century. La Lande, for example, recognized that in China and Japan paper was made not from rags but from plants such as bamboo.[15] He also distinguished that this tradition was "of great antiquity," and likely the origins for Europe's own industry. In his discussion of Japanese papermaking, La Lande cites the German naturalist Engelbert Kaempfer (1651–1716), who traveled to the Pacific and wrote *The History of Japan,* published posthumously in 1727. Kaempfer makes clear the relationship between rag shortages and an interest in non-European techniques of papermaking when he writes: "My design is only to give a short, but clear and full account of the ways of making Paper in use among the Japanese, a nation less known and less frequented, intended chiefly for the instruction and satisfaction of those, who would be willing to try the same experiment upon some barks of our European trees."[16] To aid his reader, and perhaps amateur papermakers, Kaempfer illustrated the plants employed in Japan to make paper, offering a visual aid to Europeans in search of their arboreal equivalents (figs. 21, 22).

La Lande also included in his narration of Asian papermaking techniques the Jesuit Jean-Baptiste du Halde's *Description de la Chine* (1736), in which Du Halde outlined the materials employed in Chinese papermaking:

> In the Province of Se-chwen the paper is made of hemp. . . . That in the province of *Fo kyen* it is made of soft *Bambû:* That in the Northern Provinces they make it of the Bark of the Mulberry Trees: That in the Province of *Che kyan,* it is made of wheat or rice-straw: That in the province of *Kyang nan,* they make a parchment of the skin that is found in the cods the silk-worms spin; which they call *Lo wen chi;* and which is fine, smooth, and fit for inscriptions and cartridges: In fine, that in the Province of *Hû-quang,* the tree *Chu,* of *Ko-chu,* furnishes the principal material for paper.[17]

Both the Jesuit and the naturalist focus on the variety of Asian papers. Their brief descriptions of these foreign papermaking economies celebrate the

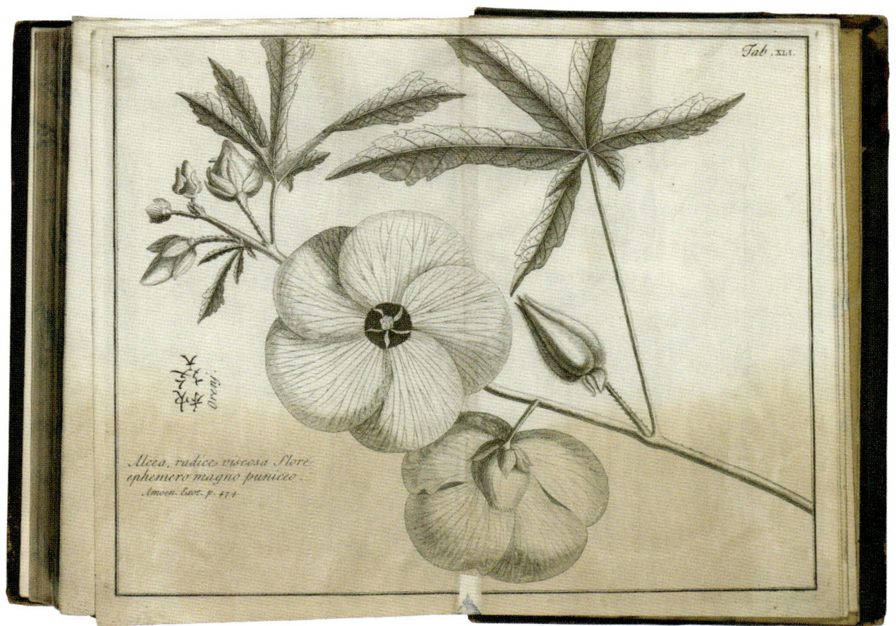

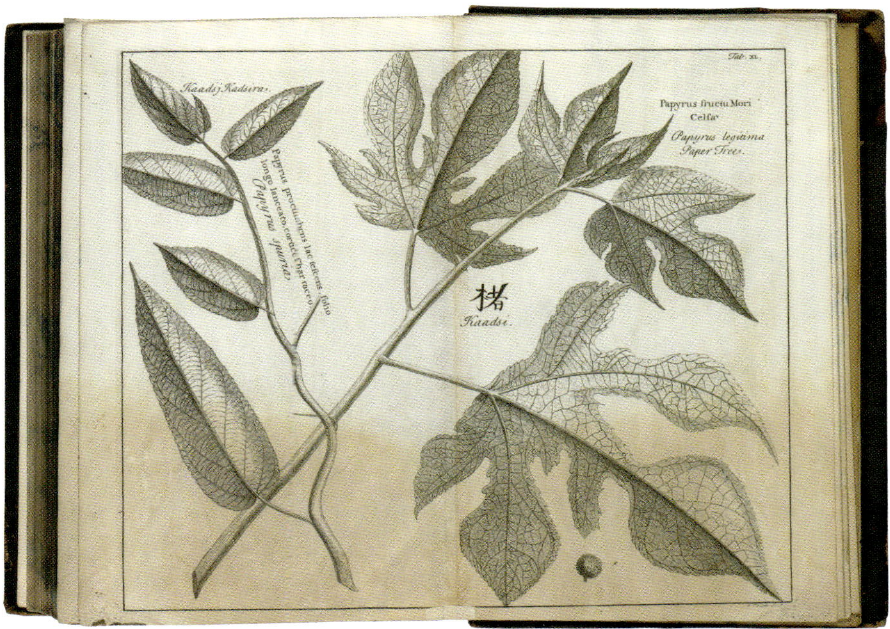

Figures 21 and 22. Top: Engelbert Kaempfer in *The History of Japan*, 1727. Engraving on paper. Houghton Library, Harvard University, Cambridge, Mass. "Alcea (Hollyhock)"; bottom: "Futo Kadsura."

regional varieties of paper and the integral relationship between local flora, fauna, and industry. Rather than categorizing a "Chinese" or "Japanese" paper, both elaborate an encyclopedia of papermaking plants, illuminating the geographic and material complexity of early modern paper outside of the European rag-paper trade. In his citation of Chinese and Japanese papermaking, La Lande gestured to two distinct and advanced paper economies, founded upon a technology that was similar to but also significantly departed from that outlined in the main body of his text. In turn, his knowledge of these distant papermaking economies depended upon print and the circulation of printed treatises produced both in Asia and Europe. For La Lande's knowledge of these worlds did not rely on a study of the paper materials themselves, but instead on the print trade produced both by Europeans who traveled to Asia and the print cultures of China and Japan that journeyed back to armchair travelers in Europe.

Encyclopedias of Paper

Du Halde, unlike Kaempfer, did not base his *Description* on his own travels but on accounts written by Jesuit missionaries in China. In his knowledge of Chinese papermaking, Du Halde possibly could have drawn from a translation of the *Tiangong Kaiwu* (*Exploitation of the Works of Nature)*, an encyclopedia written by Song Yingxing and published in 1637, which was one of the first surveys of technology in China. The compendia covered industries from hydraulic engineering to metallurgy, an influential work that would be reprinted in later treatises such as *Gujin Tushu Jicheng,* also known as the *Imperial Encyclopedia* (1725).[18] The *Tiangong Kaiwu,* considered singular in its dedication to clear explanations situated within an examination of man's place in the cosmos, was likely intended for an audience similar to the author—educated and elite. Yet it also has been suggested that Song Yingxing had in mind a wider audience of merchants and craftsmen.[19] Whereas La Lande's illustrations of papermaking existed within the two registers enumerated by Barthes and a third that abstracted the final product of paper from the technology of papermaking, Song Yingxing's are solely in the register of narrative, portraying the role of bodies and landscape in the production of paper.

Song Yingxing begins with the collection of bamboo, describing the season for cultivation, as the plants must be gathered in early summer: "The best in quality are those shoots that are about to put forth branches and leaves." He then focused on the necessary decomposition of the bamboo plant, as the stalks are broken down through an extensive process of soaking and beating: "A pit is dug right there in the mountain and filled with water in which the bamboo sticks are immersed. The water supply is constantly maintained by means of bamboo pipes so that the pit will not run dry. After soaking for more than one hundred days, the bamboos are carefully pounded and washed to remove the coarse husk and breen bark or node (this is *sha-ch'ing* or 'killing the green')."[20] In his opening woodcut, Song Yingxing places the actions of sorting and fermentation into a single plane, visually realizing the physical proximity described in the text between the gathering of the plant and soaking it on the mountain (fig. 23). The decomposed fibers are mixed with lime and cooked over an open fire for eight days (fig. 24). The pulp is washed repeatedly with clear running water, and after "some ten days [of repeated cooking, straining, and washing], the bamboo pulp will naturally become odorous and decayed."[21] Following these stages of fermentation, the pulp is placed into a tank, where the sheet will be formed (fig. 25).

Figures 23–26. Song Yingxing in *Tiangong Kaiwu*, 1637. Woodcuts on paper. Bibliothèque nationale de France, Paris. Fig. 23: "Gathering and Fermenting the Bamboo"; fig. 24: "Cooking the Bamboo"; fig. 25: "Forming the Sheet in the Paper Mold"; fig. 26: "Drying the Paper on Heated Brick Walls."

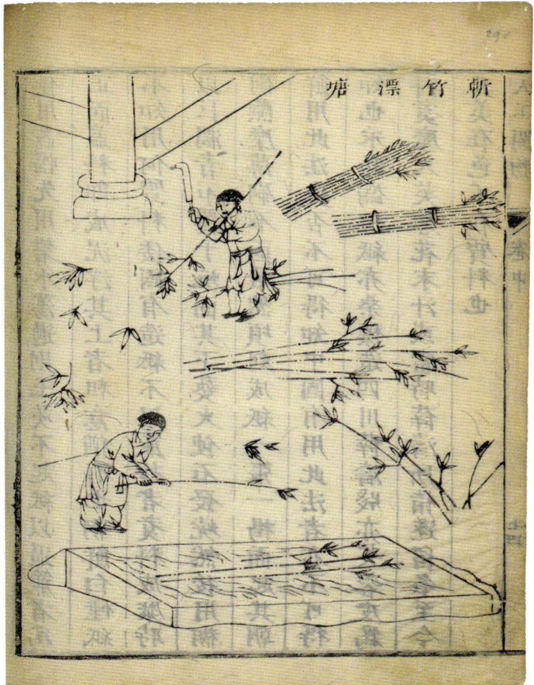

Fig. 23

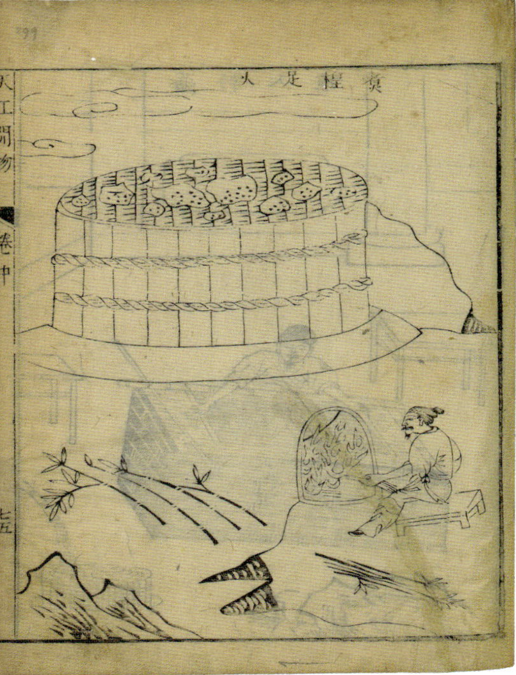

Fig. 24

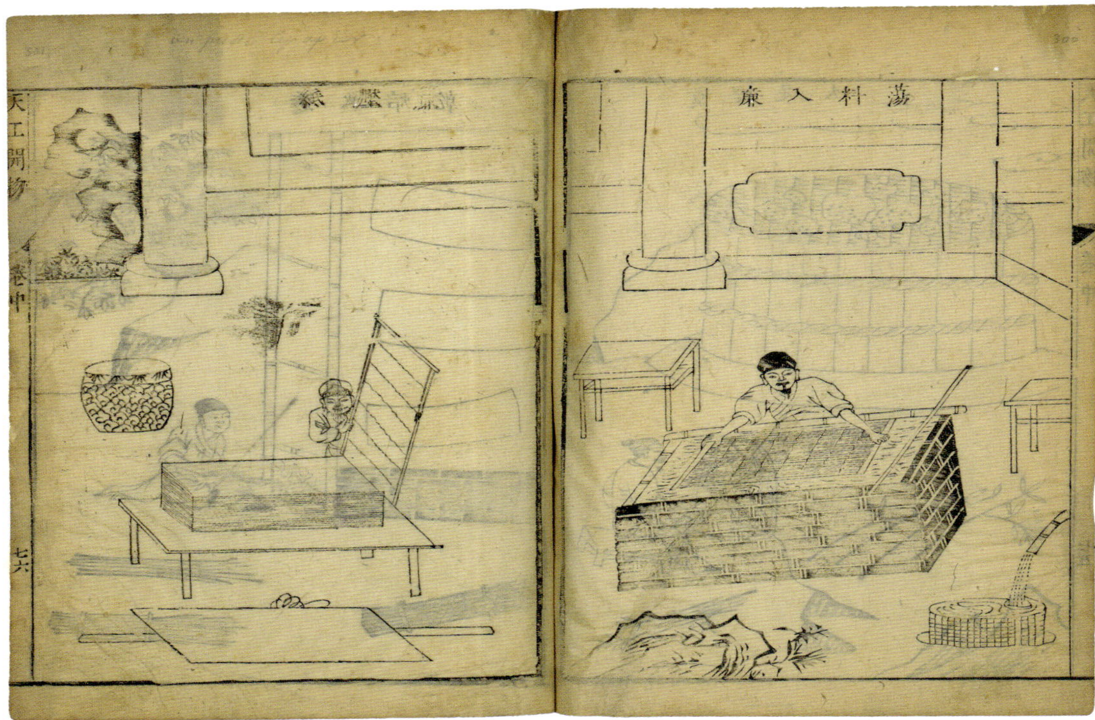

Fig. 25

Song Yingxing describes the paper mold as a "rectangular frame covered with a mat woven of finely split and polished bamboo." As in European rag paper, the action of the papermaker determines the quality of the paper: "The thickness [of the paper] will depend on the way in which the screen is manipulated: shallow submerging results in thin paper, while a deeper dip produces a thick one."[22] After formation, the sheet is removed from the mold and placed onto a wooden board: "When the number is sufficient, the sheets are covered with another board and are rope tied with the aid of a pole placed over the top board, as in a wine press, and all the water is squeezed from the sheets."[23] The paper was then placed to dry on heated bricks, as he illustrates (fig. 26).

In his explication of papermaking, Song Yingxing never isolates a single tool. Instead he integrates landscape, papermaker, process, and tools into one register. Whereas the focus on the mill's structure in La Lande's illustrations demonstrates the myriad ways in which rag paper often became invisible as it embodied other technology, Song Yingxing's woodcuts attend to the illustration of paper itself. In the final woodcuts, Song Yingxing pictures the drying sheets

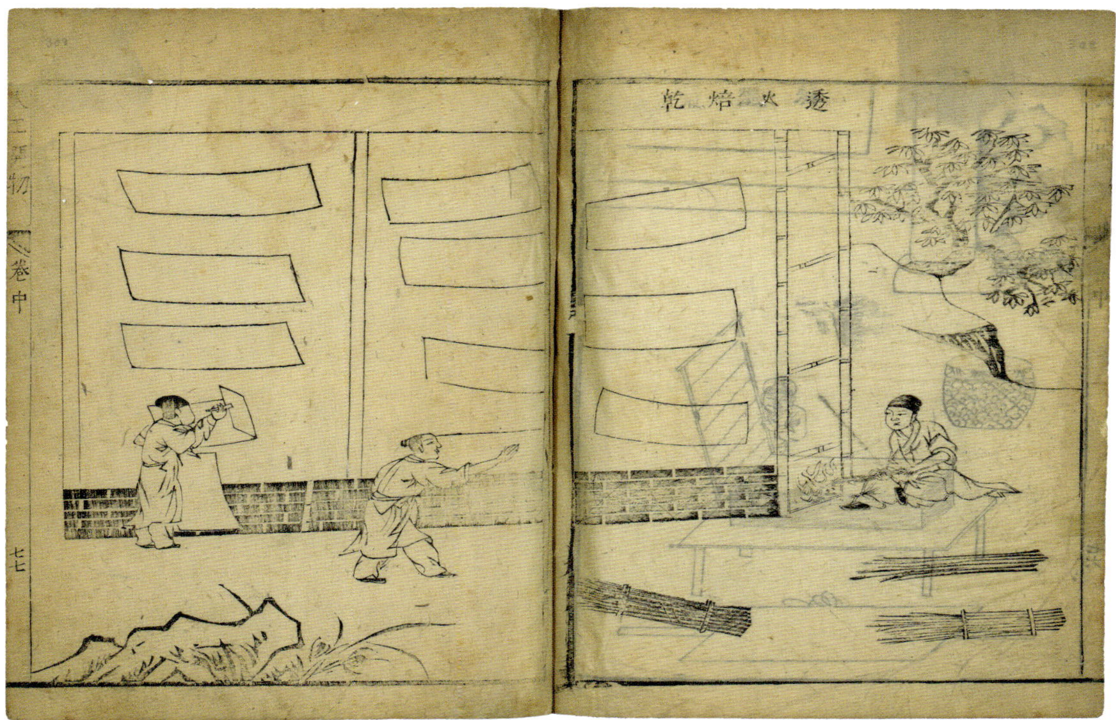

Fig. 26

against the wall. The paper of the treatise becomes the ground of the paper sheets drying—drawing a clear line of demarcation between the process under consideration and its final presence as a product in this exposition. His attention to paper's surface and presence within his encyclopedia also resonates with the extensive attention garnered to the paper support itself as a bearer of meaning in the calligraphic worlds of the Ming and Qing dynasties.

As one sixteenth-century treatise on paper illustrates, the variety of papers in early modern China far exceeded their relatively homogenous European counterparts:

> There are the: *sajn wuse fenjian,* "gold-sprinkled note-paper in five colours," of a fine but dense texture; the *wuse dalianzhi,* large-mould paper in five colours, and the *sajn zhi,* "gold-sprinkled paper." The *baijan,* "white paper" is thick and stiff like a wooden board, glossy on both sides, and as pure as jade. There also the *yinghin wuse juajian,* "gold note-paper in five colours"; the *ciqing zhi,* "porcelain-blue

paper," strong and fine like satin. . . . The *Songjiantan jian* is manufactured without starch, the *Jingzhoi lianzhi* "lian paper of Jingzhoi" is laminated, thick and glossy, because it is sized with wax, and is decorated with all sorts of bird and flower ornaments. . . . There is also the *mianzhi,* "the mulberry-tree paper," which is used to remount old painting scrolls.[24]

European paper was produced in a variety of sizes and colors, but it was predominantly celebrated as a clean, white sheet. Its chief demarcation of difference was its watermark. By contrast, the ground of the paper in China was a vehicle for meaning. As Timothy Barrett remarks, it "had a cultural value in China as part of the technology of the creative scholar that it did not have in Europe."[25] In their attention to the regional variety of Asian papers, Kaempfer and Du Halde stand on the precipice of two paper worlds. They recognize the ground of meaning accorded to paper in these non-European cultures. Yet their texts also played a role in the reception of Asian papermaking in Europe. Eventually, when the extensive papermaking economies of Asia opened to European markets, the paper produced in China and Japan lost the recognition of regional specificity and became a generalized embodiment of non-European paper.

A nineteenth-century text commissioned by the French government illuminates the reduction of global papermaking economies into a picturesque field of distant agrarian technologies. Following the Treaty of Nanking in 1842, King Louis Philippe of France sent a commercial envoy to China to open up trade and to evaluate the local commerce, including the paper industry.[26] The French diplomat in China, Marie Melchior Joseph Théodore de Lagrené (1800–1862), commissioned a series of watercolors on local industries, including one on papermaking. With the opening economy of China and portended new markets, these illustrations continue the work of *Description des arts et métiers*. In turn, the artist—who signed his name Yoeequa—cited an iconographic tradition of papermaking indebted to Song Yingxing. Yet in contrast to both La Lande's and Song Yingxing's volumes, this one contains little textual explanation and mainly consists of opaque watercolors (figs. 27, 28).[27] Unlike Song Yingxing's work, in which the craftsmen are absorbed in their work as small figures within larger landscapes, Yoeequa's papermakers stare out at the viewer, isolated in the ground of the page. With the absence of lengthy texts, the watercolors are the sole vehicle through which to convey the technology. Papermaking in China becomes a picture. Instead of producing a trade in knowledge, the watercolors create a single view that Chinese paper is different from European—made from bamboo, produced without water-mill power.

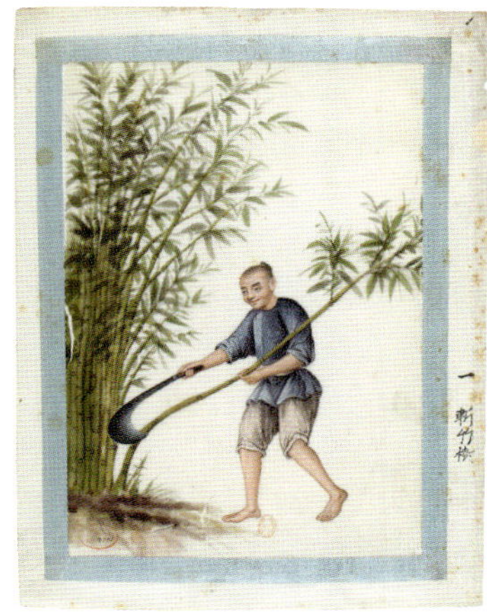
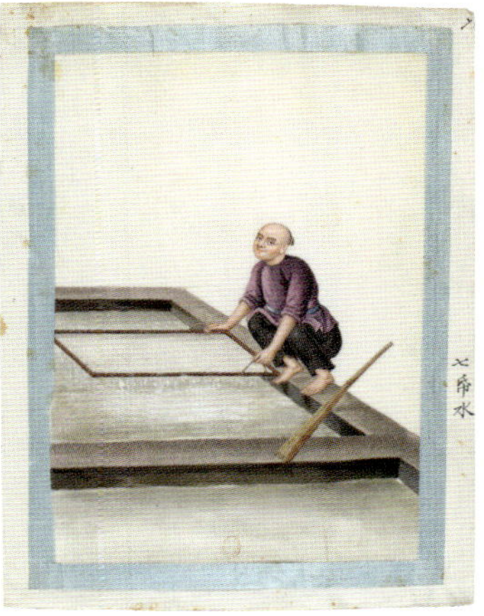
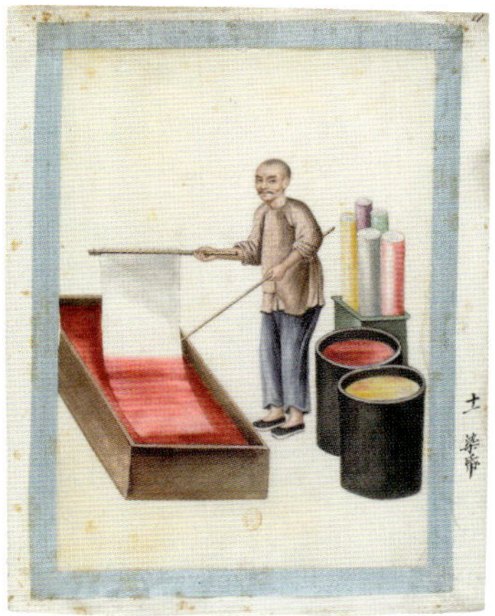

Figures 27–29. Yoeequa in *Papier: Fabrication, teinture et reliure*, ca. 1830–40. Ink on silk. Bibliothèque nationale de France, Paris. From left: "Cutting Bamboo"; "Dipping the Paper Mold"; "Tinting the Paper."

Nevertheless, in both Song Yingxing's and Yoeequa's illustrations, paper and its surfaces—as opposed to papermaking—remains the site of focus. The concluding images in Song Yingxing's work display the final product as sheets drying on the wall. In Yoeequa's work, the papermaker holds the pigmented sheet up to the viewer (fig. 29). In turn, rag paper carried meaning in its structure—its mold and watermark. This is seen in a final image in La Lande's treatise when a woman holds the sheet of paper to the light, allowing it to transmit and reveal the paper's underlying technology, from waterpower to wire mark (fig. 30).

Both Song Yingxing and La Lande were formative in creating an iconographic vocabulary for papermaking. Their treatises opened a window onto artisan worlds and expanded the knowledge of conscientious literati and philosophers about the material structure of the book. To this day, Song Yingxing's and La Lande's illustrations permeate histories of papermaking, reducing the complexity of regional variation into a single set of graphics. Moreover, this printed iconographic vocabulary for papermaking contributed to the perception of Asia and Europe as the two dominant papermaking centers. Although this was true in the eighteenth century, it elided the role of other major papermaking economies in the late-medieval world, most prominently the importance of the Islamic empires in translating the technology across Central Asia and into North Africa and Spain.

As one of the major papermaking economies of the late-medieval world, Islamic papermaking survived not in illustrated technical treatises, but in the

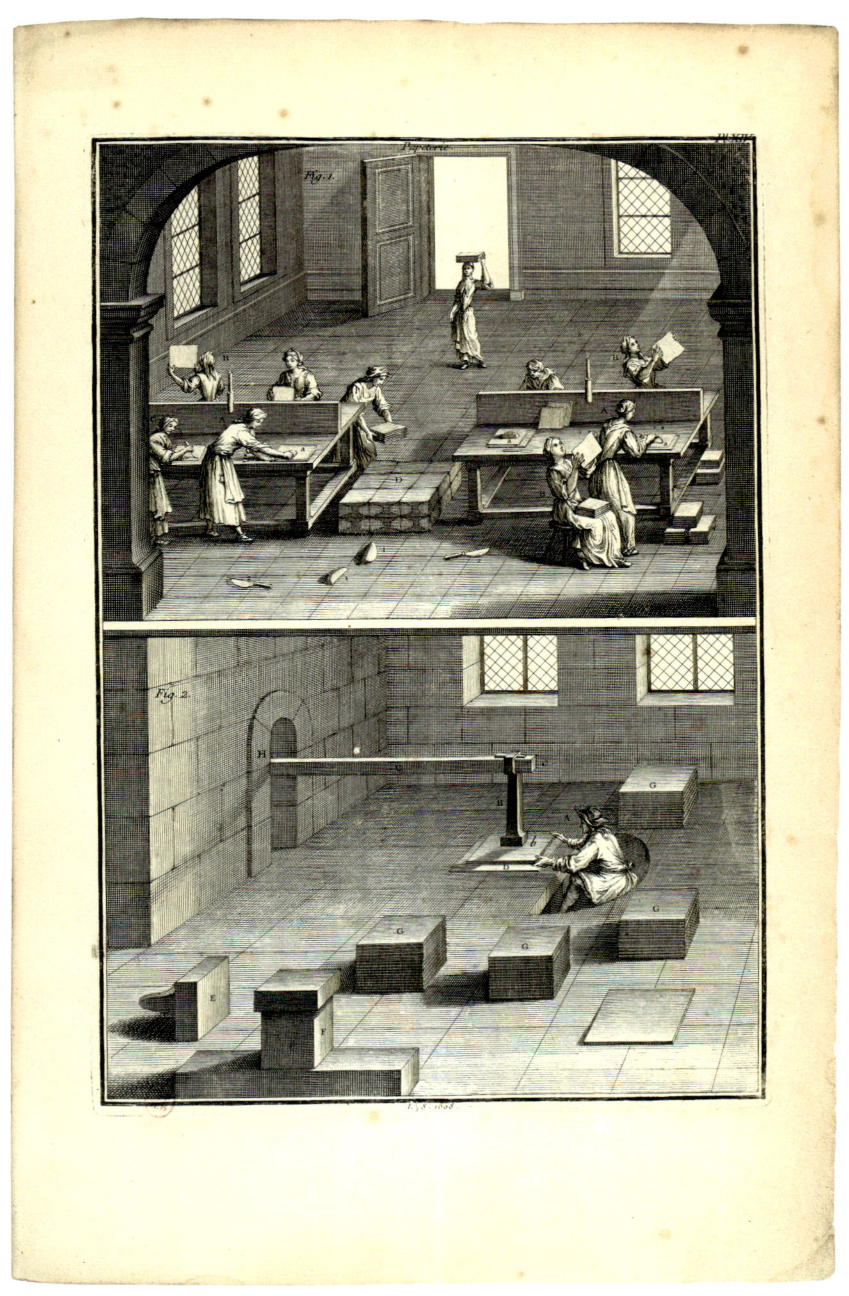

Figure 30. Jérôme de La Lande, "Packaging Paper and Quality Control," in *Art de faire le papier*, 1761. Engraving on paper. The plates are signed: "Billé sculp. 1776." Houghton Library, Harvard University, Cambridge, Mass.

paper products themselves, documents, and books that testified to an essential stage in the history of papermaking. Nevertheless, this history of paper remained forgotten, embedded in the material itself but not distilled into transmittable pictures of production. The dominant era of Islamic papermaking happened in a world before European print, and in a geographic area separated from the print culture of China and Japan. With important exceptions, writers in Arabic took little interest in devoting extensive treatises and illustrations to the history of papermaking, an absence that delayed its eighteenth-century reception in Europe, which focused on paper through its textual and visual tradition, as opposed to the presence of the paper itself.

Paper Matter

In the *Encyclopédie* (1751–66), the author of the entry on *papier,* Louis de Jaucourt, explicitly links Europe with the ingenious invention of making paper from linen and hemp: "Finally in Europe by becoming civilized, it was discovered the ingenious art of making paper with the old cloth of hemp or linen, and since this discovery, we have perfected the making of paper."[28] Jaucourt notes the production of other types, from the indigenous paper made in the Americas to the variety from China and Japan. Nevertheless, he repeatedly refers to rag paper as European, narrating an evolution from papyrus (called Egyptian paper) to parchment, cotton paper, bark paper, and, finally, paper made from linen.[29]

Throughout the eighteenth century in Europe, historians remained uncertain about where rag paper was invented, although most were certain that the technology originated on the European continent. Jaucourt suggests several possible inventors: the Germans, the Italians, and Greek refugees in Basel after the fall of Constantinople. And he mentions in passing Humphrey Prideaux (1648–1724), who suggested that rag paper came from "the Orient" because there were many Arabic manuscripts written on paper.[30] Nevertheless, the *Encyclopédie* remains equivocal, demurring from supporting Prideaux's claims, despite the archival evidence.

Yet Prideaux, the English churchman and author of *The True Nature of the Imposture Fully Display'd in the Life of Mahomet* (1697), was one of the first scholars to recognize that rag paper originated outside of Europe. In *The Old and New Testament Connected in the History of the Jews and Neighboring Nations* (1716–18), he maintained that paper was invented during the Abbasid period in the eighth century CE, and traveled into Spain from North Africa. For his evidence, he cites "the old manuscripts in Arabic, and the other oriental languages which we have from thence, are written in this sort of paper; and some of them are certainly much ancienter, than any of the times here mentioned in this matter."[31]

Prideaux, an early Orientalist, studied at Oxford, where he must have worked with the Arabic manuscripts in the Bodleian Library, and his thesis was likely informed by exposure to its expanding collection under the donations of William Laud (fig. 31), John Selden, John Greaves (fig. 32), Edward Pococke, and Robert Huntington (fig. 33).[32] Although he became one of the first seventeenth-century European scholars to note that there were many outstanding exemplars of Islamic calligraphic works on paper well before it was a common material in the Western medieval scriptorium, his work remained marginalized, and scholars such as Jaucourt refrained from endorsing Prideaux's position. Yet before Prideaux's work with Arabic manuscripts in the Bodleian, there were many opportunities to realize the lineage between the Judeo-Arabic world and paper. For the circulation of Arabic, Hebrew, and Aramaic manuscripts on paper only increased after the fall of Constantinople in 1453, the sack of Tunis in 1535, the Battle of Lepanto in 1571, and the Ottoman siege of Vienna in 1529 and in 1638. Nevertheless, Europe underwent a collective amnesia about paper's genesis.[33]

On the one hand, the source for this collective forgetting may be attributed to the invention of print in fifteenth-century Germany. On the other, the erasure of paper's origins also may be ascribed to the European theologians and scholars who established a predominantly Christian worldview. The history of paper is inseparable from religion, as demonstrated by the earliest histories of paper, which feature in surveys of religious texts. To take just one other eighteenth-century example, in *Conformité des coutumes des Indiens orientaux avec celles des Juifs et des autres Peuples de l'Antiquité* (1703; *Conformity of the Customs of the Oriental Indians with Those of the Jews and Other Peoples of Antiquity*), the author figures the history of media as essential to his study of both "Oriental Indians" and Judaism. Central to the author's exploration of religious practices are the varying supports on which cultures inscribed their sacred texts, beginning with the leaf of the Latanier tree in India and expanding to include serpent skins, brass tablets, wax-covered wooden tablets, the bones of sheep and camels, papyrus in Egypt, and parchment in Pergamon.[34] The author concludes with paper, "which is one of the most useful and commodious things that was ever found out by man."[35] Yet like his contemporaries, the narrator is unsure about paper's origins.[36]

It was not until historians were faced with the microscopic and chemical analysis of Archduke Rainer Ferdinand of Austria's papyrus collection—a trove of one hundred thousand documents on papyrus and paper discovered in

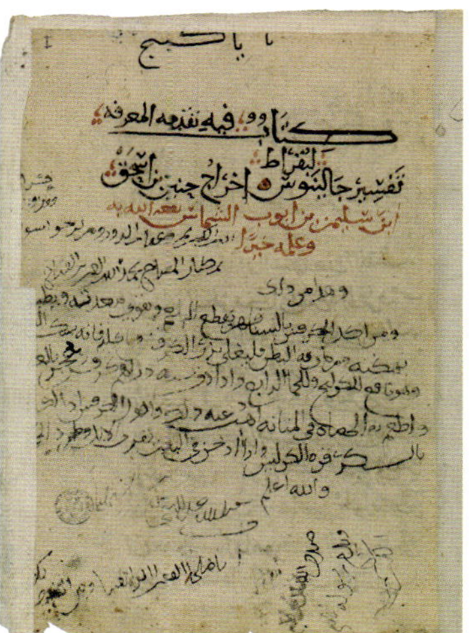

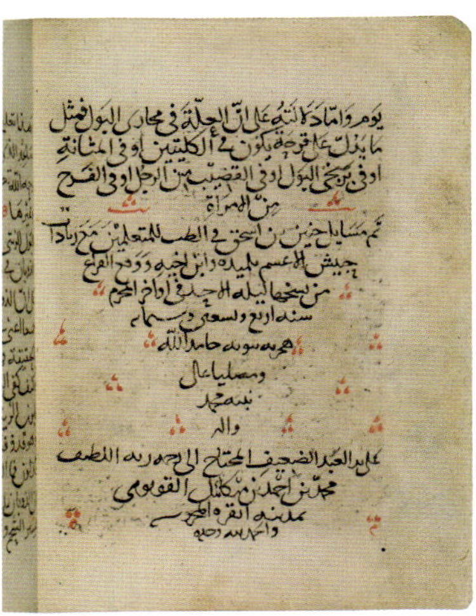

Figure 31. Fol. 1a in *Commentary of Galen on the Book of Hippocrates on Prognostics*, 13th century. Ink on paper, inscribed with the expression "o buttercup," referring to the ability of buttercups to repel the worms and insects that eat paper. Bodleian Library, University of Oxford, Ms. Laud Or. 139.

Figure 32. Fol. 85b in *Questions on Medicine for Beginners*, 1294. Ink on paper. Bodleian Library, University of Oxford, Ms. Greaves 25.

Egypt—that the invention of rag paper in the Islamic Empire was accepted. Josef von Karabacek (1845–1918) and Julius Wiesner (1838–1916) examined the paper documents and discovered that they were made from hemp and linen rags, demonstrating that rag paper was already in circulation in Arabic-speaking territories from the eighth century CE.[37] They therefore concluded that rag paper was invented in China, and that the technology migrated to the Islamic world.[38] The plant material that formed the basis of papermaking across Asia from China and into Korea and Japan was not available in Central Asia. Papermakers in Samarkand therefore innovated and began to use linen and hempen textile fragments. Since this discovery, historians of paper have traced the large-scale production of rag paper to Samarkand and the papermaking centers that emerged in Damascus, Baghdad, and Cairo. It was this tradition of papermaking that traveled into Catalonia to become the foundation of Europe's own rag-paper revolution.

Yet there are still many questions and uncertainties that remain about the technology and history of paper in this diverse region.[39] Terminology is a problem, although it is agreed that rag paper was invented in Samarkand and spread to other centers of the Islamic caliphate. Rag paper would become essential to both the bookmaking tradition of Islamic manuscripts and the bureaucratic paperwork of chancelleries. But it was also used by the disparate religious communities from Samarkand to Cairo, incorporated by not only Islamic but also Jewish and Christian communities (among others). The languages inscribed onto this paper are equally diverse, and so scholars demarcate the production

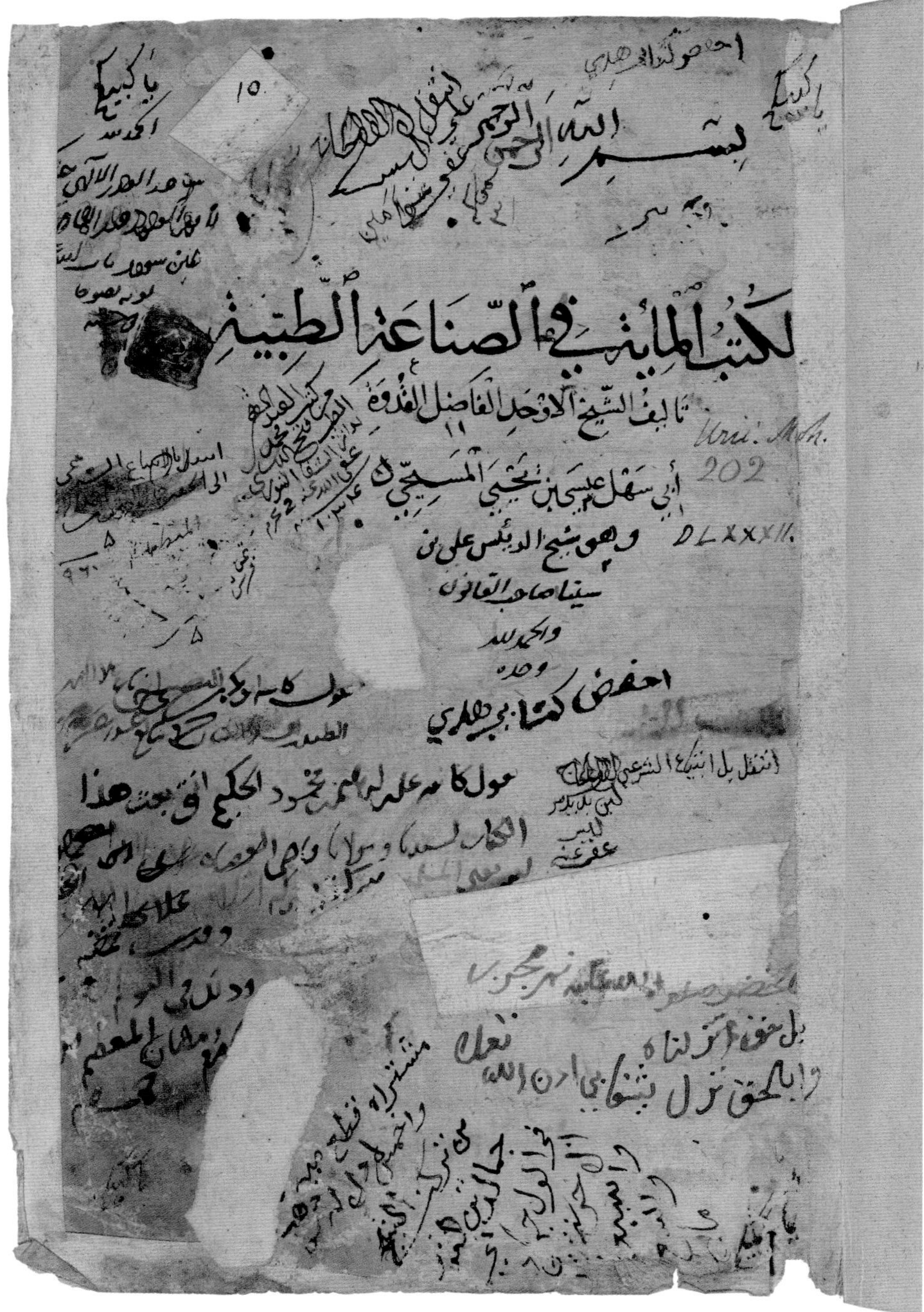

Figure 33. Fol. 2a in *The Hundred Books on the Medical Art*, 1196. Ink on paper with the inscription "o buttercup." Bodleian Library, University of Oxford, Ms. Huntington 202.

with terms such as "Arab paper," "Arab-Persian paper," and "Islamic paper," which offer necessary terminology. Yet paper historians also acknowledge the shortcomings of these markers to describe the variety of paper over a long temporal period and wide geographic area.

The study of paper technology between the eighth and thirteenth centuries in Central Asia and the Middle East also remains limited because the European paper trade supplanted the papermaking capitals in Damascus and Cairo (and elsewhere) in the fifteenth century. Many centers of this industry therefore disappeared, and the only remaining evidence is the manuscripts themselves and two surviving texts on papermaking.[40] Questions remain about the employment of water-mill power for the making of paper, although, as Jonathan Bloom points out, there is no reason to think that paper was not made with mill power.[41]

Paper was a mobile commodity that circulated, which also makes it difficult to locate and classify its characteristics from different geographic regions. Nevertheless, the rag papermaking that began in Samarkand is often distinguished by several key factors. First, it is recognized in comparison to the later paper produced in Europe because it does not contain watermarks, which were invented in the thirteenth century in Italy.[42] Therefore, in the absence of watermarks to date and geographically locate certain paper stocks, historians measure the distances between laid and chain lines to identify shared characteristics among groups of paper.[43] They also define specific regional papers based on other factors including texture, surface, transparency, color, firmness, and suppleness. The addition of starch was integral to papermaking in the Islamic world, producing a surface receptive to calligraphy brushes. Starch from a variety of plants, such as white sorghum, wheat, and rice, was often added directly to the paper pulp to increase its viscosity and aid sheet formation. Sizing—applying a ground to the paper so that it absorbs ink without bleeding—fluctuates widely according to geographic areas and the needs of various writing instruments and inks, from the brush to the quill. Due to the employment of different writing instruments, Europeans stopped using starch to size their paper and invented gelatin sizing, which made paper's surface amenable to writing quills.[44]

In later seventeenth- and eighteenth-century catalogues of Eastern manuscripts, the support is rarely mentioned. As the historian of the watermark Charles-Moïse Briquet dryly remarks, the cataloguing of "Oriental" manuscripts in the major European libraries, from the Vatican in the seventeenth century to the Escorial in the eighteenth, is exceedingly imprecise.[45] Frequently the cataloguer neglects to mention if the work is on parchment or paper, and often the term "cotton paper" (*bombycina*) remained in circulation to describe manuscripts deemed "Oriental," whereas rag paper (or *chartaceus*) was reserved for Western Latin manuscripts.[46] This stands in contrast to fifteenth-century inventories of libraries, which often explicitly cite the material of the book, as in the 1496 inventory of Gianfrancesco Gonzaga's library, in which the scribe attends to whether the book was made from *carta bona* (expensive rag paper), *carta de papiro* (papyrus), or *carta membrana* (parchment).[47] A tendency to ignore the support and its specificity remains pervasive today. In the publication of his nearly twenty-year study of the surviving thirteen editions of the *Maqamat al-Hariri,* an Islamic work frequently cited as an influence on late-medieval European art, Oleg Grabar never mentions whether the works are on parchment or paper (fig. 34).[48] Perhaps he took it for granted that the manuscripts were on paper; whereas for a Western medievalist, it would be extraordinary to encounter a deluxe thirteenth-century illustrated manuscript on rag paper.

Paper carries its own history in its body, and it was the investigations of scientists such as Karabacek and Wiesner that relayed this important chapter in its history. Yet the study of the material itself—in the examination of Islamic manuscripts—remains restricted to those who have access to the original materials, and also the necessary resources to undertake technical analysis. Although these papers proliferated in European archives and libraries, they were not accompanied by a textual or visual tradition that presented accompanying evidence. The generation of a papermaking iconography that emerged in the seventeenth and eighteenth centuries for both Asian and European papers—however limited—created a shared vocabulary that allowed for cross-cultural comparison, while also absenting the Islamic paper capitals of the late-medieval world.

The delayed reception of the material evidence also illustrates a limit in art history, founded on the photographic studies of images. The importance of photography to the discipline of art history became canonized in the slide lecture as a didactic tool, but may also be seen in more idiosyncratic works, such as Aby Warburg's *Mnemosyne Atlas* (fig. 35).[49] In many ways, the photographic archive supports art history, allowing access to materials that are otherwise unavailable. Yet this chapter in the history of paper reveals the limits of an iconography devoid of material. For it can also create a cultural amnesia about how technologies inform images, and how the commodities on which societies choose to archive themselves for the future structure knowledge-making practices.

Paper beyond the Mold

In contrast to the medieval Islamic world, from which only a few descriptions of papermaking exist, no known accounts survive in Mesoamerica from before the sixteenth century. Although some of the earliest paper fragments are from

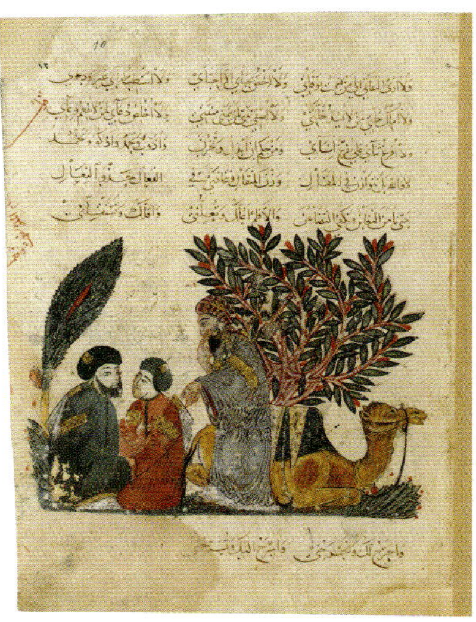

Figure 34. Yaḥyā ibn Maḥmūd al-Wāsiṭī, Fol. 138v in *Maqamat al-Hariri*, 1237. Watercolor and opaque paint on paper. Bibliothèque nationale de France, Paris.

Mesoamerica, little remaining evidence of this expansive paper world now exists. In turn, many of the papermaking accounts were written by missionaries in the Americas on the imported European paper support. This tradition continued into the twentieth century, as anthropologists and ethnographers wrote the histories of Mesoamerican papermaking, largely by studying contemporary papermaking in the communities of the Otomí to better understand its historical predecessor.[50]

These paper ethnologists also often argue that the commodity produced in Mesoamerica is not actually paper. In one of the foremost histories of Aztec and Maya paper, Victor Wolfgang von Hagen stated that although the Maya developed *huun* paper and the Aztec *amatl,* it was not paper because it was not made by placing a mold into a liquid substance with suspended fibers, the defining characteristic of paper in Asia and Europe.[51] Instead, papermakers employed wooden mallets and stones carved with geometric patters to hammer the pulp into a tensile surface. This absence of a paper mold has led historians to determine it as a secondary product to Asian and European papers. Yet Dard Hunter noted that in "both the mould-formed sheets and those beaten into leaves the raw material is the same and the finished product is similar, the only difference lying in the actual procedure used to convert the vegetable fibre into sheets or strips of paper."[52] Indeed, the Spanish conquistadores who arrived in the New World categorized the indigenous support as paper, perceiving its direct relationship to the rag paper produced in Europe, despite the divergent materials and technology.

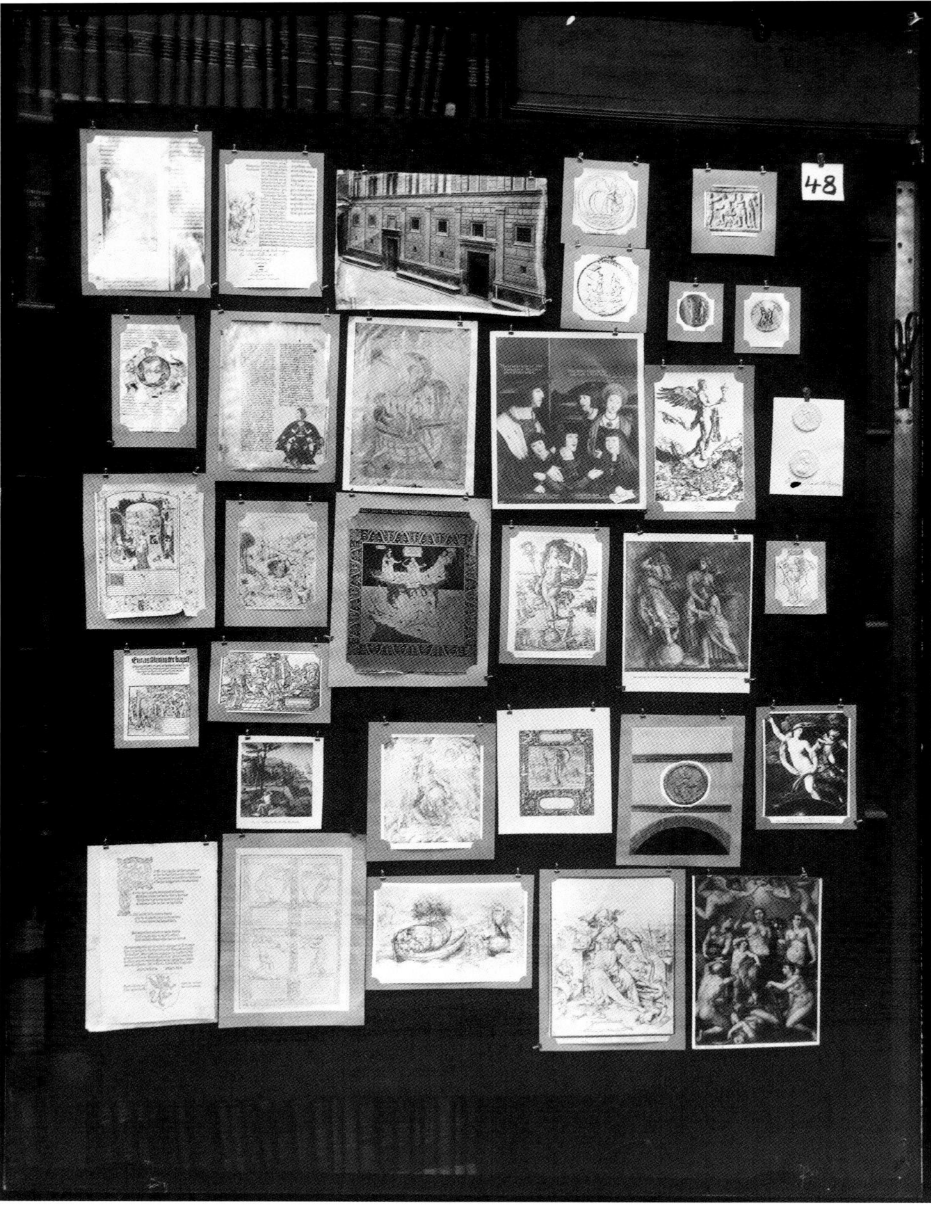

Figure 35. Aby Warburg, Panel 48: "Fortuna," in the *Mnemosyne Atlas*, 1924–29. Wooden board covered with black cloth and black-and-white photographs, 59$^{1}/_{16}$ × 78¾ in. (150 × 200 cm). Warburg Institute, London.

The earliest European accounts of the New World reveal a fascination with indigenous paper. In 1562 in the town of Maní, the Maya libraries were burned in an *auto-da-fé* (public penance under Inquisition) led by Fray Diego de Landa (1524–1579). At the same time, Landa wrote his own history of the Maya, *Relación de las Cosas de Yucatán* (1566; *The Relationship of the Things of the Yucatan*), in which he notes that "they made this paper from the roots of a tree, and gave it a white gloss upon which it was easy to write."[53] It was this application of a paste that attracted many European commentators in their descriptions of indigenous paper. Peter Martyr d'Anghiera (1499–1562), who published some of the earliest European accounts of the New World, also defined indigenous paper in his *De Orbe Novo* (1530; *Decades of the New World*) according to its prepared ground:

> The pages on which the natives write are made of the thin bark of trees, of the quality found in the first, outer layer. . . . These membranes are smeared with a tough bitumen, after which they are limbered and given the desired form; they are stretched out at will and when they are hardened, a kind of plaster or analogous substance is spread over them. . . . One may write thereon whatever comes into one's mind, a sponge or a cloth sufficing to rub it out, after which the tablet may be again used. . . . It is supposed that the natives preserve in these books their laws, the ritual of their sacrifices and ceremonies, astronomical observations, and the precepts of agriculture.[54]

For Peter Martyr, indigenous paper was distinct for its capacity for reuse. His description resonates with a type of European writing surface: the erasable tablet, which was prepared with a ground receptive to a metalpoint stylus.[55] These tablets proliferated in everyday transactions, particularly among merchants, as the surface could be easily effaced and reused, offering an economical tool for short-term record-keeping and accounts. The tablet was not, however, a surface on which humanists, artists, and theologians archived their work for posterity. It is noteworthy that the conquerors who sought to eradicate indigenous culture from the landscape primarily described the local writing material in regard to its capacity for erasure and rewriting.

The limits of the Spaniards' histories and descriptions must be taken into account, for the surviving papers from Mesoamerica introduce a variety that diverges from the white-paste paper remarked upon so often. Some legal documents, such as the Oztoticpac Lands Map, are not recorded on a white-paste ground (fig. 36). The map illustrates the litigation over inalienable and

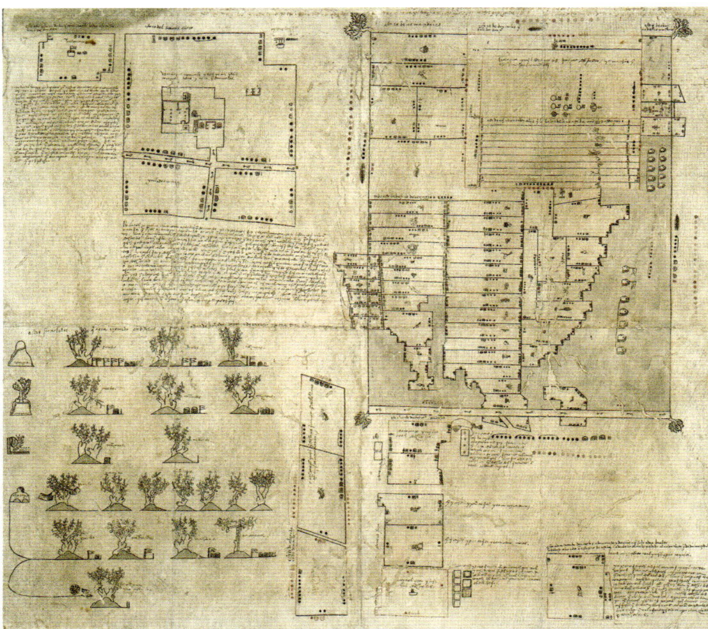

Figure 36. The Oztoticpac Lands Map, ca. 1540. Red and black ink on amatl. Library of Congress, Geography and Map Division, Washington, D.C.

transferable property in Texcoco circa 1540, and the paper contains no preparatory ground layer and instead features its own intricate patterning of fibers.[56] Without the starch paste, this fibrous support remains resistant to eradication, challenging later alterations to the drawn borders.

The limited accounts of Spanish missionaries and conquistadores on papermaking in New Spain are complemented by twentieth-century descriptions written by paper historians and ethnographers. This paper ethnography inadvertently suggests that little changed in the technology from the production of the earliest paper fragments in the fourth century CE to the twentieth century. To study the technology of a particular twentieth-century papermaking community, in order to historicize nearly two thousand years of papermaking across an entire continent, places the technology of Mesoamerican papermaking within the mythology of a homogenous culture unchanged by time. Yet the papermaker Dard Hunter's study of Otomí papermaking also illuminates the process of making paper without a mold.

In Hunter's narration, the Otomís gather the bark of the fig trees in "the autumn when full of sap. After the bark is well dried it is placed in a pool of running water, which washes away the parenchyma or glutinous substance, leaving the pure fibres." Like the paper produced elsewhere, the process relies upon the extensive use of running water to macerate and ferment the fibers. And similar to papermaking practices throughout the world, lime—in the form of ash—is incorporated into the final stages of creating the pulp. As Hunter reports, the bark fibers are "laid in a stream where the material receives a further cleansing. It is then boiled with ashes, or in the liquid in which corn tortillas have been boiled. A large earthen pot of native construction, heated over an open fire, is used in boiling the bark."

The process only departs from those previously described in the actual formation of the sheet, in which "the fibres are beaten with wooden clubs or mallets until they have separated and are in a pulpy condition. When the material has been thoroughly macerated it is made into a paste and spread evenly over a board in a thin sheet with the fingers; and then gently beaten with a small stone, which mats the fibres, forming a homogenous sheet of paper."[57] Hunter's description of twentieth-century papermaking only offers a glimpse into the papermaking world of Mesoamerica pre-1500. The samples of paper from the Otomí that Hunter included in his publication have little in common with the paper from the codices that survive today, as Hunter's fragments are coarser and more fibrous. Moreover, Mesoamerican paper carried the same variety as papers in China or Japan, depending on the local landscape. Technical studies of Mesoamerican paper suggest that paper made in the cold regions was from the leaf of the maguey plant, that the temperate regions used the bark of the yucca (*izote*), and that in hotter areas used that of the fig tree (*amate de tierra caliente*).[58]

One of the few surviving pre-conquest indigenous representations of the papermaking economy is in the *Matrícula de Tributos* (sixteenth century; *Tribute Roll*), a document consisting of sixteen folios on amatl recording the tributes brought to Tenochtitlan by conquered territories (figs. 37, 38).[59] In folios 3v and 4r of the *Matrícula de Tributos* for the provinces of Cuauhnáhuac and Huaxtepec, paper is represented in its scrolled form as a tribute brought to the capital (figs. 39, 40). The document was likely painted a few years before the arrival of the Spanish, and it has attained a prominence in studies of the codex because of its many translations. Most famously, the *Matrícula de Tributos* was copied by a Nahua scribe to form one third of the Codex Mendoza, a work on European paper documenting various aspects of Aztec civilization for a European audience (figs. 41, 42). When the original document was copied, the scribe reformatted the order of the tributes for a reader accustomed to reading left to right. The original document also received an extensive gloss in Spanish, transforming it into a "Rosetta Stone" for the study of the Aztec.[60]

Yet when the *Matrícula de Tributos* was copied onto European paper and formatted into a codex with other texts, the integral role of amatl as both a sup-

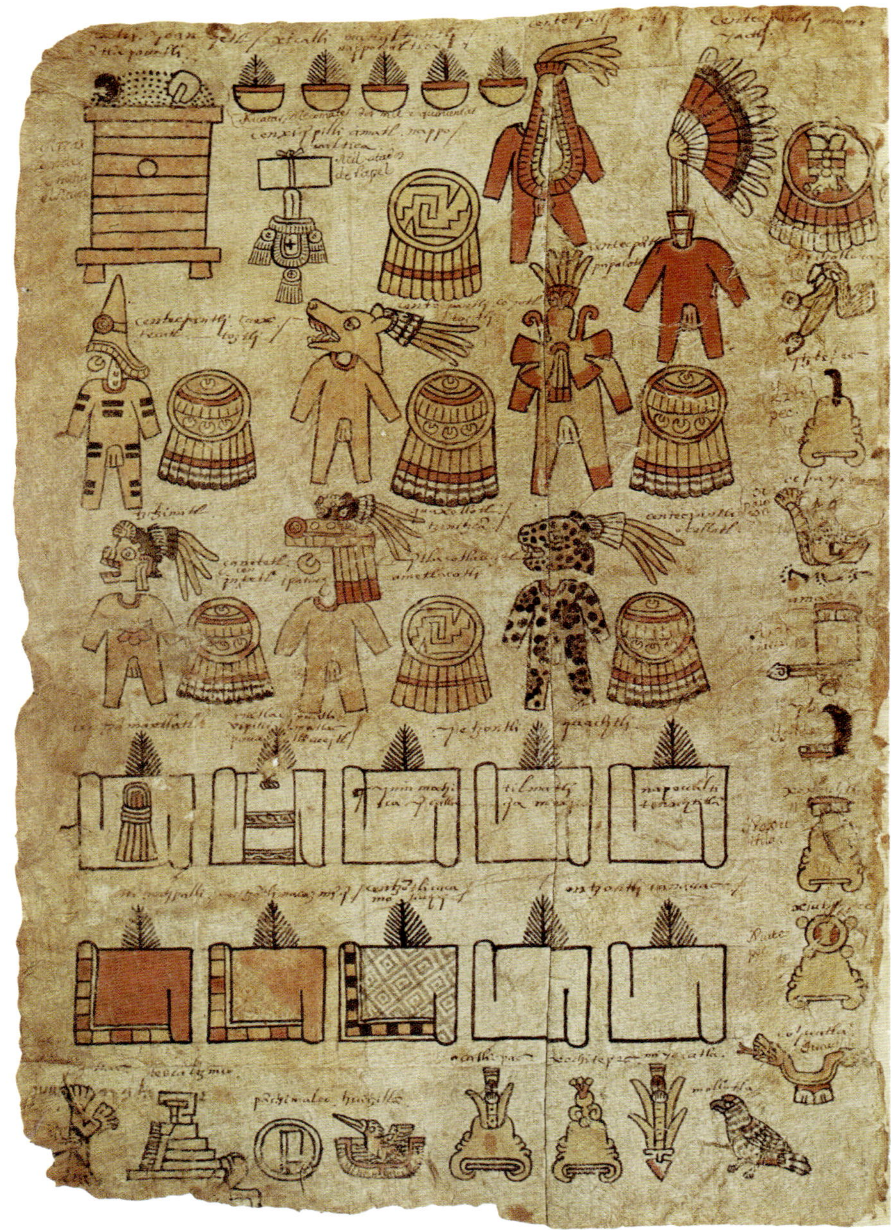

Fig. 37

Figures 37–40. *Matrícula de Tributos*, ca. 1520. Ink and paint on amatl. Instituto Nacional de Antropología e Historia, Mexico City. Fig. 37: fol. 3v; fig. 38: fol. 4r; fig. 39: fol. 3v (detail); fig. 40: fol. 4r (detail).

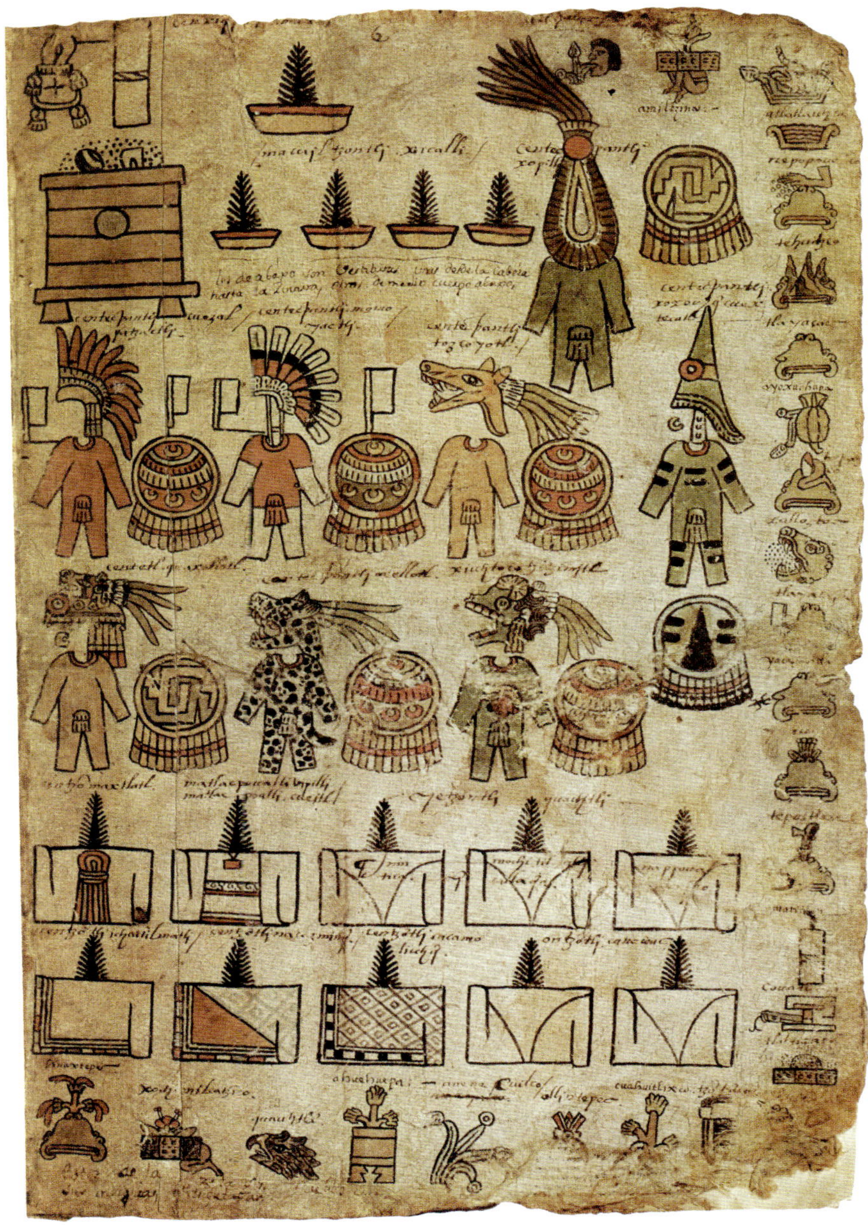

Fig. 38

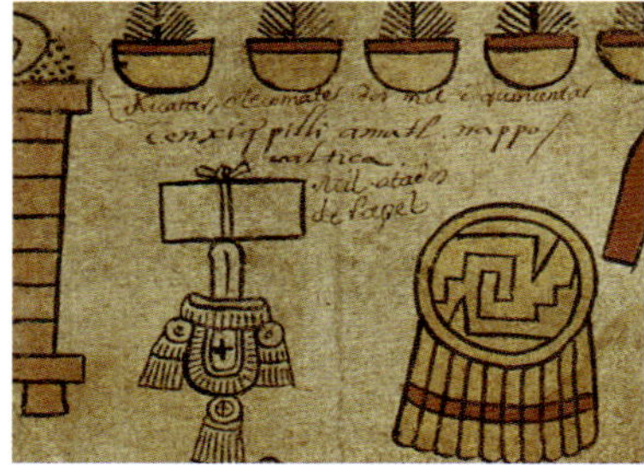

Fig. 39

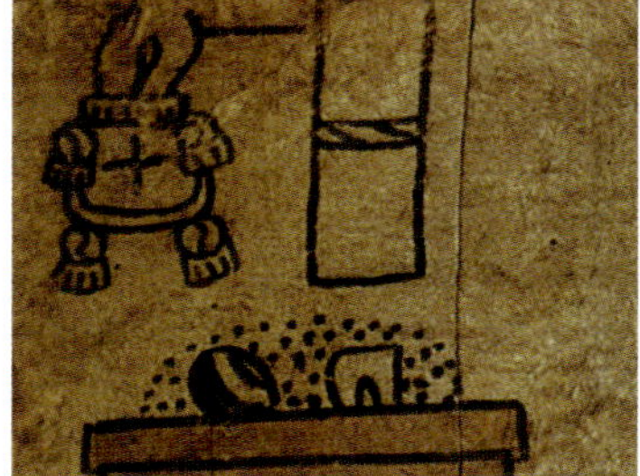

Fig. 40

port and a commodity was lost. In later copies, print took part in obfuscating previous material histories, as when Samuel Purchas published sections of the Codex Mendoza in his *Hakluytus Posthumus: or, Purchas His Pilgrimes* (1625). Whereas the Codex Mendoza reoriented the placement of the tributes on the page, Purchas divided the commodities into discrete sections, framed by lengthy typographic exposition (fig. 43). Purchas singled out the representation of paper twice, making the pictograph legible to a European audience, clarifying that the single scroll stands in for "eight thousand sheets of paper."[61]

In his employment of the term "sheets of paper," Purchas evokes the flat, stacked reams of paper produced in Europe. Yet as the glyph makes clear, this paper rolled and unrolled, and was not the flat-sheet reams shipped across the Atlantic from Genoa to Mexico City. For a European audience, the glyph of paper would evoke a chancellery document, as the hanging stone from the scroll resonates with the large wax seals that bound certified European documents and charters on both parchment and paper. The image of the scroll also reverberated with the history of papyrus scrolls across the Mediterranean, pointing toward an earlier world of *papyri,* as discussed by Pliny in his *Naturalis historia.* When Pliny recounts the existence of a letter written on papyrus by Sarpedon from Troy that was recently rediscovered in a temple, he doubts the authenticity of this document. For, as he rhetorically asks, if papyrus existed in the time of Homer, then, "Why, too, has Homer stated that in Lycia tablets were given to Bellerophon to carry, and not a papyrus letter?"[62] As the quip makes clear, papyrus was valued for the same qualities that would come to define European paper: its lightness, mobility, and ability to transverse vast distances. Many Europeans commented on the notable lightness of indigenous paper, and the pictorialization of it in the form of the scroll illustrates its ability to be transported across long distances with relative ease. Yet local accounts of paper in New Spain also suggest that paper carried a certain weight, both material and metaphysical.

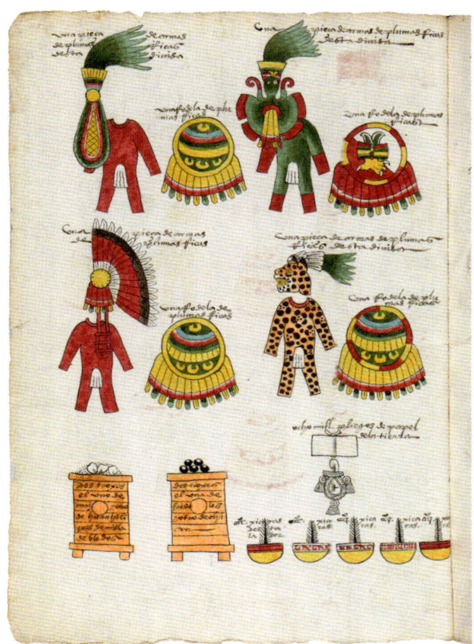
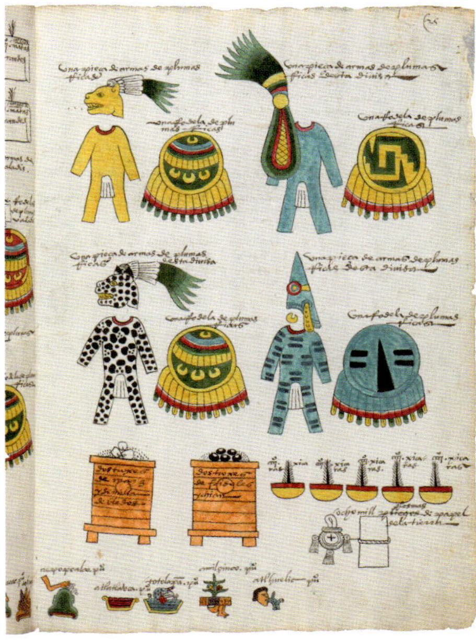

Figures 41 and 42. "Tributes from the Province of Quauhnahuac" in the Codex Mendoza, ca. 1541. Paint and ink on European paper. Bodleian Library, University of Oxford, Ms. Arch. Selden. A. 1. Left: fol. 23v; right: fol. 25r.

European and Asian paper sizes always were determined by the length and width of the paper mold's frame, which was restricted by the arm span of the papermaker, who had to maneuver the mold. Yet paper made without a mold does not bear this restriction. The absence of a mold allowed papermakers to create paper in unprecedented and unregulated sizes. As the Franciscan Friar and proto-ethnographer Bernardino de Sahagún (1499–1590) remarked about the paper offerings for sacred ceremonies: "And this was a paper, white paper and not yellow paper, a finger thick, a fathom wide, and twenty fathoms long. With ceremonial arrows, arrows hardened in fire, they supported it; they were made only that they might carry [the paper]."[63] Paper is not in the discrete, regulated sizes of the European support, but an offering with weight. The difficult transport of paper—made burdensome by its incredible length and heft—determined its status as an offering. It was supported by "ceremonial arrows" made for the sole purpose of conveying this precious commodity. This singular description should not apply to all paper across the Americas. The employment of paper from divine offering to administrative material illustrates its variable

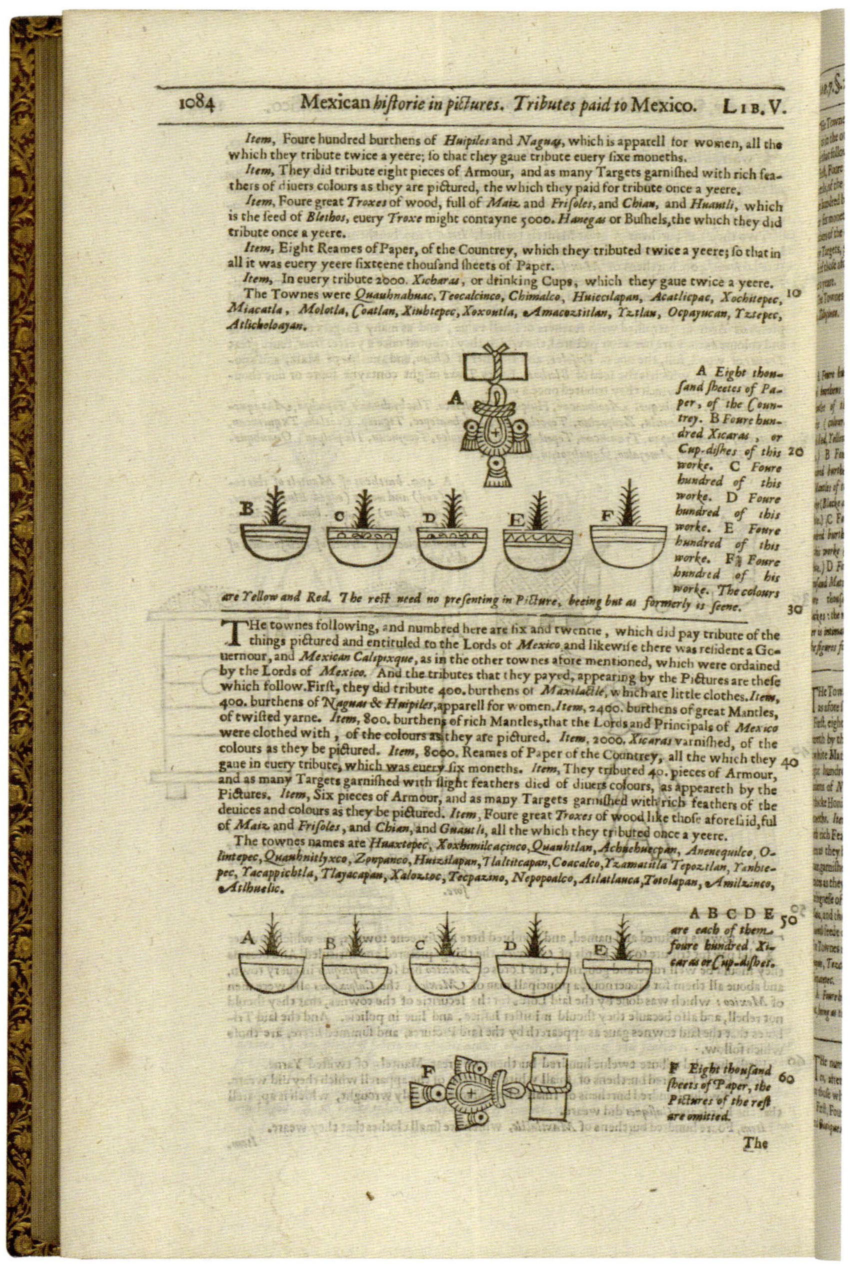

1084 Mexican *historie in pictures. Tributes paid to* Mexico. Lib. V.

Item, Foure hundred burthens of *Huipiles* and *Naguas*, which is apparell for women, all the which they tribute twice a yeere; so that they gaue tribute euery sixe moneths.

Item, They did tribute eight pieces of Armour, and as many Targets garnished with rich feathers of diuers colours as they are pictured, the which they paid for tribute once a yeere.

Item, Foure great *Troxes* of wood, full of *Maiz* and *Frisoles*, and *Chian*, and *Huautli*, which is the seed of *Blethos*, euery *Troxe* might contayne 5000. *Hanegas* or Bushels, the which they did tribute once a yeere.

Item, Eight Reames of Paper, of the Countrey, which they tributed twice a yeere; so that in all it was euery yeere sixteene thousand sheets of Paper.

Item, In euery tribute 2000. *Xicharas*, or drinking Cups, which they gaue twice a yeere.

The Townes were *Quauhnahuac, Teocalcinco, Chimalco, Huiecilapan, Acatlicpac, Xochitepec, Miacatla, Molotla, Coatlan, Xiuhtepec, Xoxoutla, Amacoztitlan, Yztlan, Ocpayucan, Yztepec, Atlicholoayan.* 10

A Eight thousand sheetes of Paper, of the Countrey. B Foure hundred Xicaras, or Cup-dishes of this worke. C Foure hundred of this worke. D Foure hundred of this worke. E Foure hundred of this worke. F Foure hundred of his worke. The colours are Yellow and Red. The rest need no presenting in Picture, beeing but as formerly is seene. 20 30

The townes following, and numbred here are six and twentie, which did pay tribute of the things pictured and entituled to the Lords of *Mexico* and likewise there was resident a Gouernour, and *Mexican Calpixque*, as in the other townes afore mentioned, which were ordained by the Lords of *Mexico*. And the tributes that they payed, appearing by the Pictures are these which follow. First, they did tribute 400. burthens of *Maxilactle*, which are little clothes. *Item*, 400. burthens of *Naguas & Huipiles*, apparell for women. *Item*, 2400. burthens of great Mantles, of twisted yarne. *Item*, 800. burthens of rich Mantles, that the Lords and Principals of *Mexico* were clothed with, of the colours as they are pictured. *Item*, 2000. *Xicaras* varnished, of the colours as they be pictured. *Item*, 8000. Reames of Paper of the Countrey, all the which they gaue in euery tribute, which was euery six moneths. *Item*, They tributed 40. pieces of Armour, and as many Targets garnished with slight feathers died of diuers colours, as appeareth by the Pictures. *Item*, Six pieces of Armour, and as many Targets garnished with rich feathers of the deuices and colours as they be pictured. *Item*, Foure great *Troxes* of wood like those aforesaid, ful of *Maiz* and *Frisoles*, and *Chian*, and *Guautli*, all the which they tributed once a yeere. 40

The townes names are *Huaxtepec, Xoxhimilcaçinco, Quauhtlan, Achuehuecpan, Anenequilco, Olintepec, Quauhnitlyxco, Zonpanco, Huizilapan, Tlaltitcapan, Coacalco, Yzamatitla Tepoztlan, Yauhtepec, Yacappichtla, Tlayacapan, Xaloztoc, Tecpazino, Nepopoalco, Atlatlauca, Totolapan, Amilzinco, Atlhuelic.*

A B C D E are each of them foure hundred Xicaras or Cup-dishes. 50

F Eight thousand sheets of Paper, the Pictures of the rest are omitted. 60

The

Figure 43. Samuel Purchas, "Reproduction of the Codex Mendoza," in *Purchas His Pilgrimes, in Five Bookes*, 1625. Woodcut on paper. Houghton Library, Harvard University, Cambridge, Mass.

uses. Yet the brief evocation of paper as an offering with mass reveals a paper distinct from the mobile, light, and economical commodity in Europe.

In turn, paper was important as both an offering and an adornment, a history that is stressed in the work of Sahagún, who describes paper in regards to its use during sacred ceremonies.[64] This continues in the literature on indigenous papermaking today, which frequently situates it in relationship to a metaphysics of destruction and renewal.[65] As the scholar Philip Arnold argues, amatl "was created out of the destruction and recombination of various forms of life, and, as a material object, it articulated a human connection to the land."[66] The writings and descriptions of paper's employment in religious ceremonies began with early European ethnographers, such as Sahagún, and continued in the work of Frederick Starr, who studied the ritual use of paper among the Otomí. Yet these accounts also place Mesoamerican paper within the domain of the sacred, obscuring the bureaucratic and bibliographic culture that paper also supported, and which Europeans systematically attempted to destroy. Paper in Mesoamerica therefore remains "primitive," "religious," and distinct from the purported secular Enlightenment technological revolution that would define European rag paper in the eighteenth century, as articulated by La Lande's treatise and his celebration of mill power.

Nevertheless, paper in Europe also was entrenched within a Christian worldview, as illustrated in one of the earliest representations of European papermaking, *Das Ständebuch* (1568; *The Book of Trades*), comprised of emblematic poems by Hans Sachs and woodcut illustrations by Jost Amman. The book begins with the pope and progresses to the cardinals, bishops, priests, monks, and beggars. After this ecclesiastical realm, Sachs and Amman introduce earthly power, beginning with the emperor. The first professions are doctor, apothecary, astronomer, and procurator, then comes print media. In the hierarchy of this book of trades, the proximate placement of print professions relative to law and medicine illustrates the importance of printed media for the structure of society. Sachs and Amman divide print into the following professional categories: typecaster, printer, woodblock cutter, papermaker (fig. 44), engraver (fig. 45), painter of prints, and bookbinder. In the world of *Das Ständebuch,* the papermaker is embedded within both a Christian hierarchy and an economy of print, making papermaking inseparable from the other professions of type cutter, bookbinder, and printer. Paper was vital to the emergence of print, but the book of trades also demonstrates how quickly paper became inseparable from the printed dissemination of texts and images, while it was disconnected from other scribal histories. In the seventeenth century, papermaking in Asia was

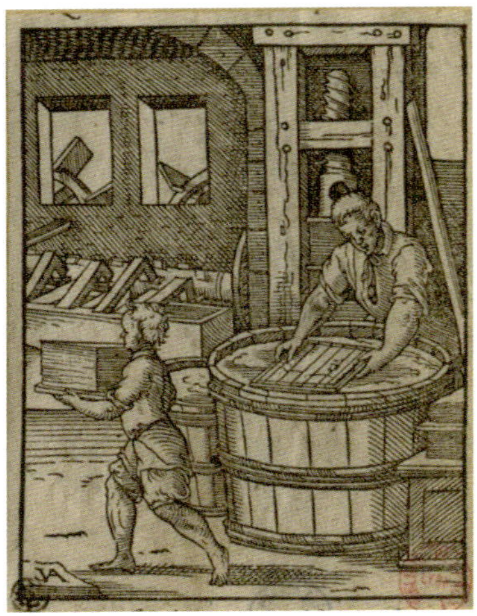

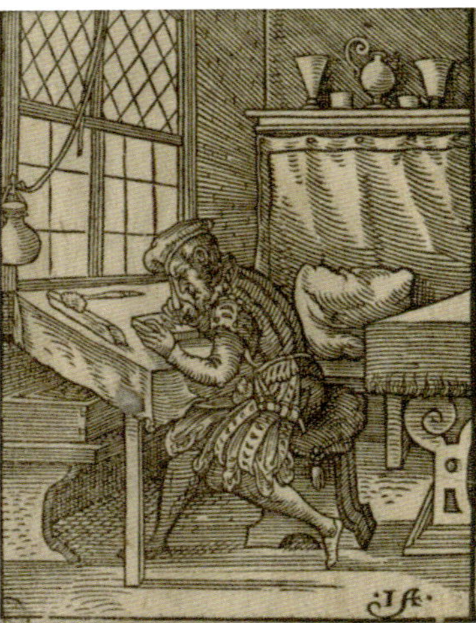

Figures 44 and 45. Jost Amman in Das Ständebuch, 1568. Woodcut on paper. Bibliothèque municipale de Lyon. Far left: "Papermaker"; Left: "Engraver."

presented as a corollary to this history of paper and print in the Christianized West. China and Japan were as renowned for their innovations in print as their production of paper, and the importance of the Asian print and paper economy was in parallel to its development in Europe.

In turn, the paper histories of two of the dominant papermaking economies before 1300—Mesoamerica and the Islamic Empire—were never quite forgotten, but they also never achieved the prominence accorded to European and Asian papers in the seventeenth and eighteenth centuries. A primary factor (among many) in the near-erasure of two central papermaking worlds was Gutenberg and typography. Both Europe and Asia produced not only paper, but also printed texts and images that widely disseminated both technologies. By contrast, paper in Mesoamerica and the Islamic empires served scribes. This manuscript culture, in turn, was translated by European publishers and printers into printed treatises—as may be seen in Purchas's publication of the Codex Mendoza.

Similarly, in the sixteenth century, Arabic manuscripts began to circulate in print and on European paper. In 1514 Pope Julius II had a book printed in Arabic on the times of prayer (fig. 46). In the 1530s Paganino da Paginini produced a version of the Qu'ran in Arabic in Venice. In 1583 Jakob Mylius published an Arabic grammar with a translation of the Epistle to the Galatians. And in the late sixteenth and the early seventeenth century, the Medici press issued many volumes in Arabic, including the Gospels with illustrations.[67] With the

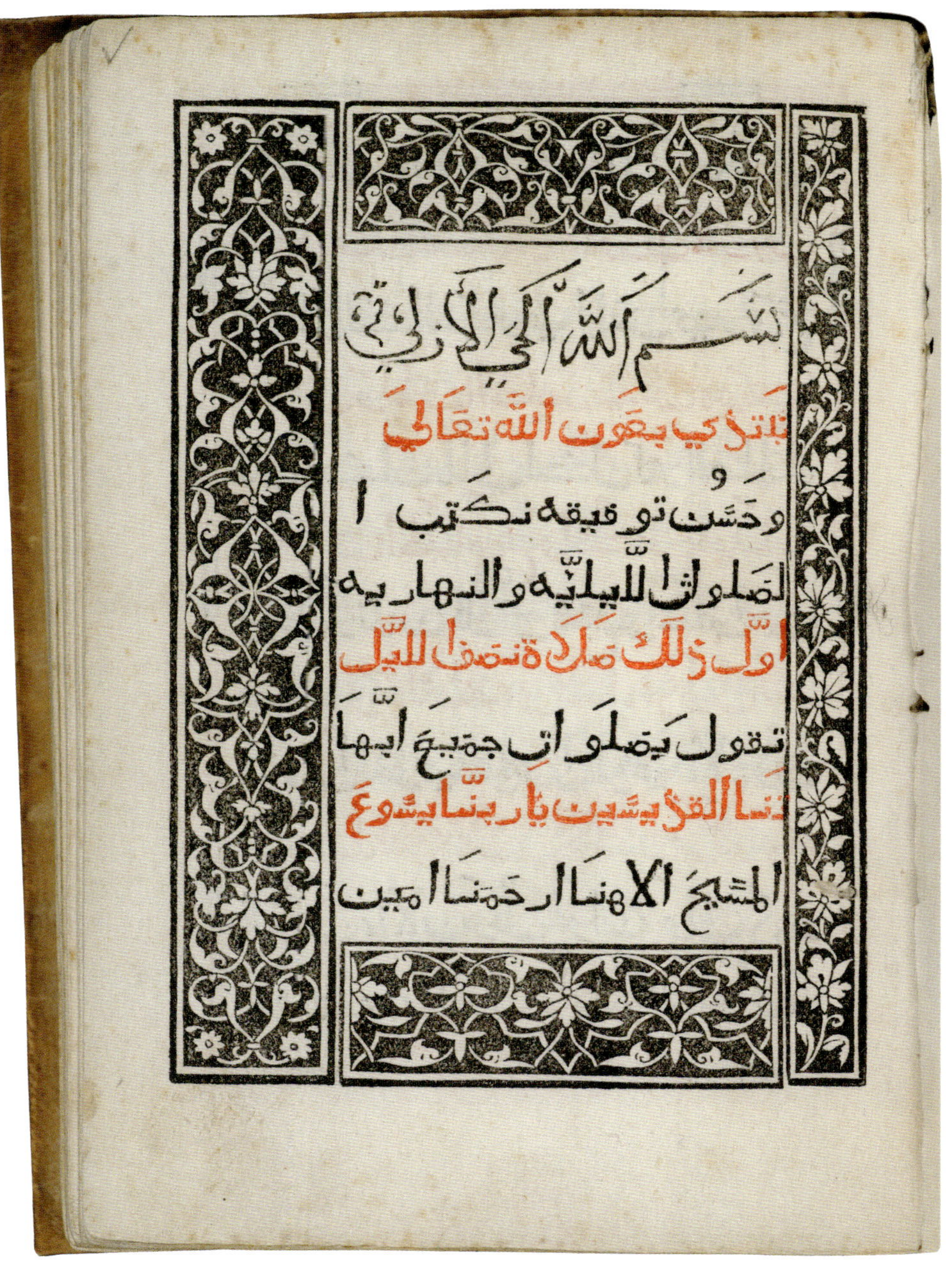
بسم الله الحي الازلي
نبتدي بعون الله تعالى
وحسن توفيقه نكتب ا
الصلوات الليليه والنهاريه
اول ذلك صلاة نصف الليل
تقول بصلوات جميع ابها
تنا القديسين يا ربنا يسوع
المسيح الاهنا ارحمنا امين

Figure 46. *Kitāb ṣalāt al-sawā'ī*, 1514. First printed book in Arabic. Rare Books and Special Collections, Princeton University Library, Princeton, N.J.

proliferation of print, the technology embedded in the earlier production of Islamic, Judaic, and Mesoamerican manuscripts was nearly forgotten. As Christian scholars studied languages and distant cultures as a means to better understand Christianity, this knowledge was distilled into a specific form dominated by European production: the printed codex on paper impressed with the European watermark, ready for export.

CHAPTER II

Forgetting Paper's Origins

Paper is vile. In his treatise *Adversus Iudeorum inveteratam duritiem* (twelfth century; *Against the Inveterate Obduracy of the Jews*), Peter the Venerable Abbot of Cluny (ca. 1092–1156) writes about rag paper, which he specifically associates with Judaism, and the senselessness of the Talmud. As he writes: "'God,' you say, 'reads the book of the Talmud in heaven.' But what type of book is this?" The abbot enumerates different forms of supports for books, including papyrus and parchment, and then he mentions books "made from the scrapings of old cloths or perhaps even some more vile material, and it is written by quills or marsh reeds and stained with some ink." As the theologian expounds: "How could he who gave the Law to the Jews written on stone tablets with his finger—that is, by the Spirit—learn anything from the book of the Jews?"[1] The support made from linen and hempen rags expresses his excoriating hatred for Judaism. Paper is explicitly non-Christian. And it is this diatribe against Judaic manuscripts that historians invariably cite as one of the earliest discussions of paper in Europe.[2]

As the abbot's digression illustrates, paper's history is weighted with centuries of Christian intolerance toward Judaism and Islam. A late-medieval history of paper also draws a map of trade and power, in which Europe is at the periphery. Janet Abu-Lughod articulates this view in her work on the thirteenth-century global trade economy.[3] In just one example, Abu-Lughod demonstrates that although Venice and Genoa were active in thirteenth-century Middle Eastern trade, their subsidiary role is revealed by their use of gold coin from Constantinople or Egypt.[4] The places of paper production mirror currency, and it is not a coincidence that Venice and Genoa also would become two centers of early modern paper manufacture, as they were two of the first ports to actively import paper. These two early sites of import would become centers for export as papermaking technology took hold in Europe. Genoese paper, in particular, was the commodity that dominated the market in New Spain. Tracing itineraries of late-medieval paper reveals Europe's movement from the periphery to the center in trade, as paper was the material on which commerce was conducted, recorded, and administered.

Judeo-Arabic Paper and Mediterranean Trade

One of the most significant bodies of evidence for the pre-European paper trade is the Cairo Genizah, a trove of some three hundred thousand manuscript fragments (most on paper) found in the attic of a synagogue in Cairo. According to an established practice, Jewish communities would not destroy a surface bearing the word of God, but instead set it in a designated place known as the Genizah. This collection of texts gathered by a Jewish community and placed in the attic of the Ben Ezra Synagogue from the ninth through the nineteenth century contains fragments in Arabic, Hebrew, and Aramaic of legal documents, letters, poetry, and religious texts. The contracts and correspondence among merchants attest to the economic relationships between the Mediterranean and the Indian Ocean in the late-medieval period, and chart paper's itineraries.[5] For example, the documents track the exportation of Egyptian paper to Tunisia, Yemen, and India and the importation of paper from Spain and Damascus into Egypt.[6] The arrival of Italy and France's dominance in the paper trade becomes marked in the fourteenth century, as documents appear on paper with watermarks, a sign that the paper is now being imported into Egypt from Italy or France, and that European cities are gradually overtaking Baghdad and Damascus as centers of paper manufacture.

Long before Italian and French paper dominated the market, however, high-quality paper was produced by the predominantly Muslim papermakers in Valencia, particularly Játiva. When King James I conquered Valencia in 1238, he assimilated the industry into his own bureaucratic needs, which formed a turning point in the acceptance of paper in Christian Europe. Following the re-conquest of Valencia, Aragonese monarchs depended on the paper made by Muslim papermakers for their administration.[7] As Richard Burns demonstrates, paperwork and documents were vital to the governing of the kingdom, as it was the "first monumental use of paper by a European government."[8] When the predominantly Muslim paper industry came under the control of the king, the economic support encouraged extensive record-keeping, and the copying of documents onto both parchment and paper to underwrite their survival as may be seen in the *Llibre del Repartiment de Valencia* (1237–52), three volumes that archive the divisions of land and property among the conquerors.[9] Even as paper was becoming essential to governmental record-keeping, it was often met with resistance. In 1145 Roger II in Sicily ordered that all official documents on paper be copied onto parchment, to assure their preservation. In 1221 Frederick II forbade notaries to use paper. In 1236 a document in Padua states that notarial documents on paper were not valid.[10] Yet in 1235 a document in the Genoese archives names an Englishman, Gualterius Englesius, who would receive 27 *soldi* per month for making paper.[11]

Outside of the governmental networks, there was a form of paperwork traveling across the Mediterranean: the correspondence of merchants. Paper defines the political and commercial relationships of Pisa and Florence with North Africa from the twelfth to the sixteenth centuries, as it was the central support employed by non-European merchants. Most of the twelfth-century diplomatic and commercial documents on trade between Italy and the Islamic world in Arabic, Latin, and Italian are on paper. For example, the privilege granted by the sultan Al-Adil I to Pisan merchants in Egypt from 1208 is written on paper in Latin, or the list of favors granted to Pisan merchants by the same sultan in 1215–16 was written on paper in Italian.[12] There is also the letter dated to 1211 from Abd-el-Wahid, responding to the consul of Pisa, Goffredo de'Visconti. Written in Arabic and translated into Italian onto what scholars in the twentieth century called *carta orientale,* the letter promises to maintain its treaty with Pisa.[13]

In turn, the prominence of Italian paper in the fourteenth century becomes evident in a document from 1366, which promises safety to Pisan merchants. Written in Arabic and Italian, the contract's substrate signifies its European provenance with the telltale brand of the watermark.[14] In her work on paper and diplomacy, Megan Williams demonstrates the signifying potential of a single sheet of paper, illustrating the complex displays of power for which paper provided the literal structure. She describes diplomacy as "a series of contingent performances acted out, in many cases, on paper."[15] The quality and type of support for contracts that were exchanged among Mediterranean governments and merchants in the late-medieval period must not be taken for granted, as it impressed the legitimacy and power of the individuals represented in the contracts. Moreover, the appearance of Western paper for trade and commerce in the archives highlights a shift in the economic dynamics of the global economy. As Europe became an economic power, so its transactions were carried out on paper manufactured within European borders.

Notary practices in Europe further contributed to paper's success, creating demand for a product on which to draft the increasing number of documents and contracts. As historians note, this was a period in which Europe underwent a "scriptural explosion."[16] In fourteenth-century Venice, notaries were required to keep registers on rag paper and records of redacted acts on parchment.[17] More notarial paper documents survive in the archives of Venice than in any other late-medieval Italian city.[18] The notaries' early embrace of paper in Venice perhaps may be explained in part by its relationship to the city's trade routes pointing east. Similarly, Siena's use of paper for administrative purposes dates at least as far back as the first volumes containing the deliberations of the General Council in 1248.[19] Like Venice, Siena's economy also was built on an east-facing trade with Mongol-Eurasia.[20]

Late-medieval Bologna also experienced an upsurge of papermaking in conjunction with escalating bureaucracy and notaries' efforts to produce registers of contracts archived in the *Memoriali* and *Provvisori*.[21] The *Memoriali* (on parchment) recorded every contract (involving a financial transaction) to protect against fraud. In turn, the *Provvisori* documented the collection of taxes paid from registering a contract. The *Provvisori* kept two copies: one parchment, one paper. This traffic between paper and parchment for official documents also demonstrates a development in archival consciousness, in which the preservation of documents becomes increasingly vital to the self-definition of a community within the city walls.[22]

This standardization of keeping documents and legal procedure informed the sheet of paper itself. A law of measurement presented in the fourteenth-century Bologna Stone was installed for public view into the walls of the Palazzo d'Accursio, attesting to the increasingly regulated sizes. It demarcates the four sizes of official paper produced in Bologna, enshrining a statute of 1389: "We decree and order that anyone who makes or shall make or shall have made sheets of paper, is bound to make them or have them made according to the fixed sizes, which are shown on the marble plaque placed on the wall next to the Palace of the City of Elders, underneath the wooden porch to the palace, and on the same bit of wall are found other standards or measures of the city of Bologna, and these are measures for Imperial, Royal, Median and Chancery sheets."[23] In his work on *pietre di paragone* (stones in city walls to establish the system of measurement for a *commune*), Emanuele Lugli argues that this system was a means to encourage economic growth, as it both controlled commerce and protected the citizens with a transparent regulatory system.[24] In the case of the Bologna Stone, it demonstrates that sizes of paper were regulated just as the measurements for the goods discussed on the paper were controlled.

Simone Martini and Early Christian Paper

Paper provided both artists and notaries with an economic tool for provisional drafts. Yet in one of its earliest and most public appearances in Italy, it was incorporated into a monumental commission by Simone Martini, who evoked paper as a technology of the law with origins in the Judeo-Arabic world. In his *Maestà* (1315), Simone incorporated two sheets into the fresco in the council chamber of the Palazzo Pubblico in Siena (fig. 47). Paper is the material of the document held in the Christ Child's hand, standing in the divine court of the Virgin (fig. 48), and it is the support on which Jerome writes in one of the roundels surrounding the central composition (fig. 49). In the first instance, the Virgin and Child sit enthroned under a *baldacchino* encircled by saints. A text unfurls from the hand of the Christ Child with its words, *Diligite iustitiam* (Observe justice), positioned toward the viewer. Yet the Christ Child's scroll is not constructed from paint. He holds a paper document.[25] In turn, Jerome sits in his chamber and translates the Bible. Here, Simone secured paper to the wet wall, so that a sheet physically represents the surface on which Jerome writes.[26] Paper becomes the material of the Vulgate, and the incarnation of Christ.

Scholars were surprised when technical evidence revealed that Simone used paper and not parchment in his quattrocento multimedia collage. Paper was still a relatively new technology in Italy and widely considered more ephemeral than parchment. Nevertheless, the city council had been using imported paper since at least 1248, and there are extant Sienese drawings on paper from the end of the thirteenth century, once attributed to Simone.[27] Paper was, however, employed in this period for preliminary notation, or as a second copy in conjunction with the more durable parchment version. It is striking, therefore, that Simone chose paper for this monumental civic image. Yet his decision may be understood in light of paper's history. The first paper mill in Italy, established in Fabriano, was only thirty years old when Simone painted this fresco. But Italians had been importing paper from Spain and North Africa since at least the eleventh century. If Simone saw Arabic, Aramaic, or Hebrew texts, it is likely that the scripts would have been written on paper.[28]

Whether Simone's inclusion of paper within the fresco can be explained as a flight of experimental fancy or pride in a burgeoning local industry, it is significant that the prophet whom he depicts with paper was Jerome, who traces with his finger the lines of text in a book rendered in paint, while writing on a

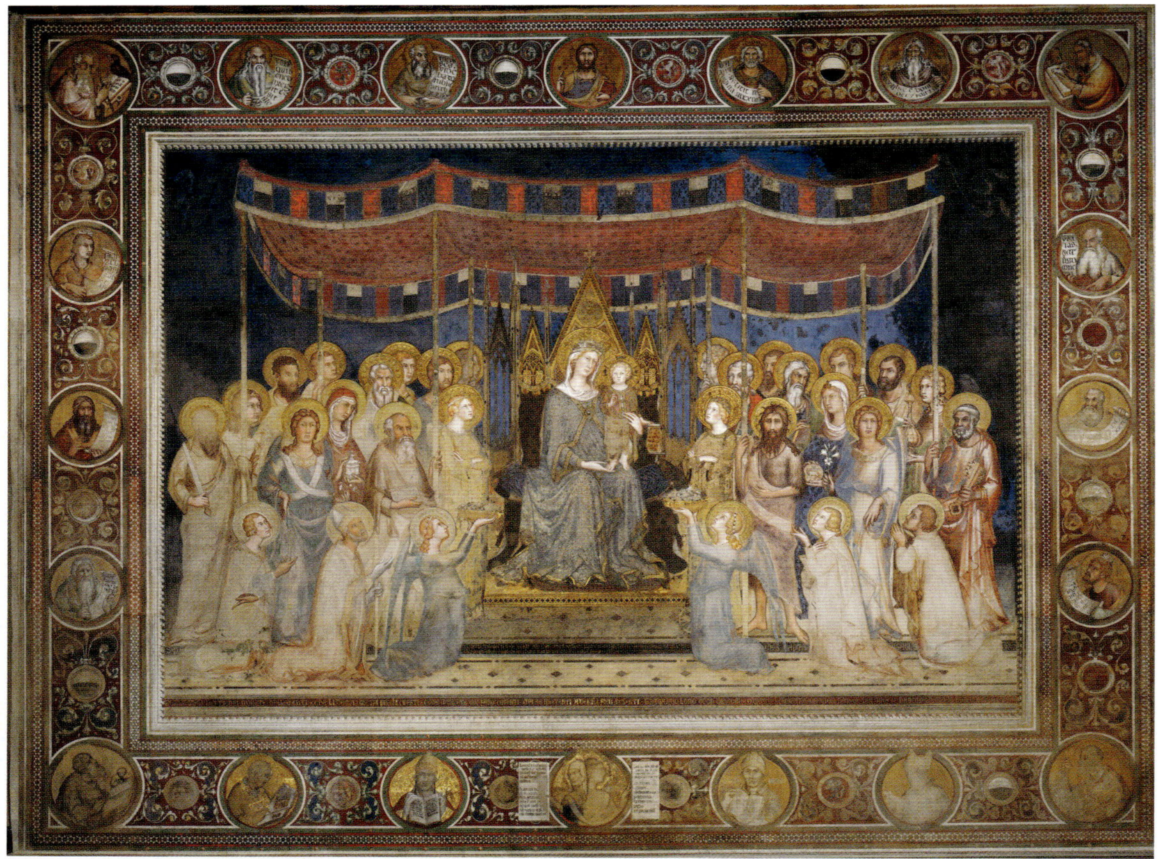

Figure 47. Simone Martini, *Maestà*, 1315. Fresco. Palazzo Pubblico, Siena.

sheet of paper. Simone envisions translation as a physical process of working across materials. The painted codex stands in contrast to a physical sheet of paper. Translation does not only occur at the linguistic level but also through the support. The attention to scribal tools suggests that texts acquire their meaning through both their content and the material supports that carry it.

As a translator of the Old Testament, Jerome was associated specifically with the languages of the Holy Land, in particular Hebrew, Aramaic, and Arabic. In his prefaces to Judith and Tobit, Jerome discusses his translation from Chaldean (Old Aramaic) into Latin. Renaissance artists noted Jerome's linguistic expertise, as he was consistently shown with Arabic pseudo-scripts from the

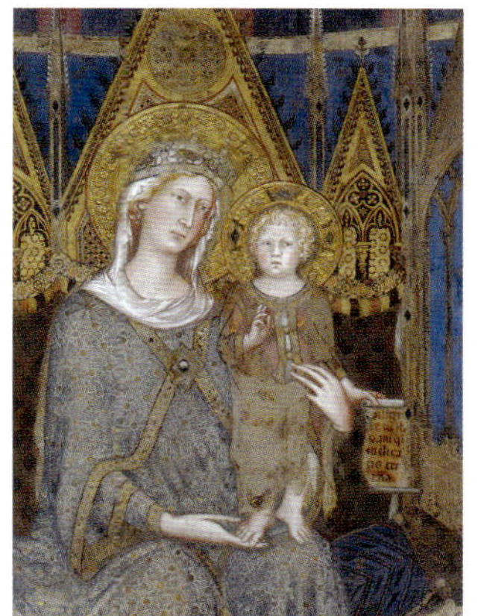

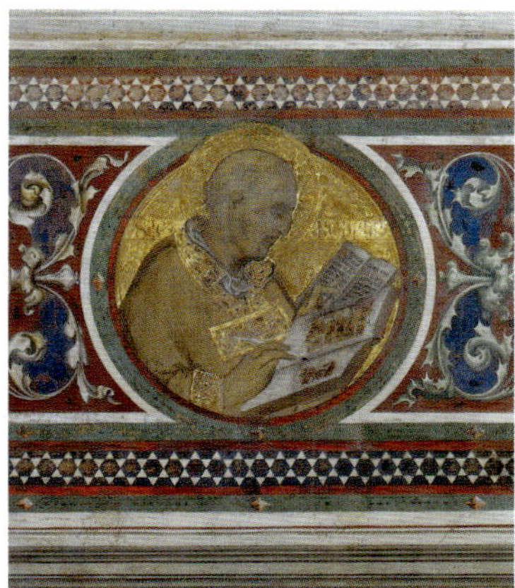

Figure 48. Simone Martini, "Christ Child with Paper Document," detail of *Maestà*.

Figure 49. Simone Martini, "Jerome with Paper Book," detail of *Maestà*.

thirteenth to the fifteenth century.[29] Moreover, Simone carefully selected the text that Jerome was inscribing on the paper, namely the preface to his translation of Job.[30] In this introduction, Jerome remarks on the problem of translation, and whether to make a direct correspondence among words, or to indirectly evoke the meaning of the original text: "And this translation follows no translator of the ancients, but will rather convey from the speech itself (which is) Hebrew and Arabic and sometimes Syrian, now words, now meanings, now both together."[31] Simone chose paper as the support for this brief meditation on translation. In this passage, Jerome also explicitly mentions the languages in which he works: Hebrew, Arabic, and Syriac. These scripts and foreign tongues arrived in Siena on paper. Although Italy was home to a new burgeoning paper industry, its origins pointed east. And it is precisely this tension between the new and the old that structures Jerome's philosophy of translation.

When Pope Damasus tasked Jerome with compiling a definitive version of the New Testament from contradictory and incomplete texts, Jerome prefaced his work with a letter in which he addressed the problem of translation: "You urge me to make a new work from the old, and that I might sit as a kind of judge over the versions of Scripture dispersed throughout the whole world, and that I might resolve which among such vary, and which of these they may be which truly agree with the Greek. Pious work, yet perilous presumption, to change the old and aging language of the world, to carry it back to infancy, for to judge others is to invite judging by all of them."[32] Jerome stated that he would reform the "aging language of the world" into the more modern tongue, Latin.

In turn, he would remove the scribal and interpretative errors that had crept into the earlier translations and return to the meaning embedded in the Greek source text. Jerome, therefore, articulated translation as a Janus-faced method. It both modernizes the text and returns it to its originary meaning. The material of paper, with its roots in the Holy Land, emblematizes this double-movement. Paper is at once a support indicative of the increasing mill-power technology in Italy and a material with origins in the Holy Land. Moreover, the physical material of paper represents the process of translation as described by Jerome: "Making a new work from the old." Paper was made from used linen and hempen fragments, and from these old rags a new substance was formed. Paper-making represents the transition from something old into something new.

Simone's deployment of paper within the fresco points toward a multivalent understanding of the material as both contemporary and ancient. Its contemporaneity expressed the modernity of Jerome's Vulgate translation in Latin. And yet paper also was the definitive support for the Arabic, Aramaic, and Hebraic texts appearing from the Holy Land. Paper was a material with ties to the geographic territories of Jerome's journeys through Asia Minor and Syria. Paper did not exist there in Jerome's lifetime, but by Simone's era it carried the connotations of the Holy Land and the territories in which Jerome had lived and worked. For the Sienese, the languages that Jerome knew, such as Hebrew, Aramaic, and Arabic, arrived on paper, suggesting that it was a commodity with origins in biblical antiquity. As Rosamond Mack contends in her work on Islam and Italian art, in the early Renaissance, people confused "contemporary eastern Mediterranean material culture with that of the Early Christian era."[33] Or as Alexander Nagel writes, "Orient was origin, our origin."[34] Italian artists considered contemporary Arabic indicative of early Christianity.[35] For Simone, paper was an invention from the Holy Land.[36] Its incorporation into his fresco materialized the imagined surfaces on which Jerome might have written. At the same time, Simone recognized paper as something new in Italy, embodying the modernization of Jerome's work.

Andrea Mantegna and Paper Diplomacy

The resonances of paper with a biblical antiquity persisted. In one prominent example, at the Council of Ferrara in 1438, the Egyptian delegates gave to Pope Eugene IV a tenth-century Egyptian codex with the Gospel of Luke in Arabic inscribed on paper (Vat.ar.18).[37] It is one of the earliest known Arabic versions of the New Testament. As a diplomatic gift, this paper manuscript from Egypt with Arabic script must have conveyed an antiquity not found in any other material.

The circulation of these types of diplomatic gifts on paper likely influenced Andrea Mantegna, who was one of the last Italian painters to explicitly evoke the origins of paper. As Nagel observes, Mantegna was particularly interested in Semitic languages, incorporating Hebrew, pseudo-scripts, and even possibly the imitation of Syriac scripts into his paintings.[38] This subtle awareness of paper's origins reveals itself in Mantegna's *Ecce Homo* (ca. 1500), a meditation on justice, legal procedure, and bureaucracy (fig. 50). The accusers of Christ wear paper crowns, while his condemnation is scrawled on paper. Whereas Simone emblazoned the words "Observe Justice" on an actual sheet, here Mantegna shifts paper into the sphere of representation. Moreover, it becomes a judicial instrument of the Passion, as both the regalia donned by torturers and the surface on which to record the verdict.[39] Mantegna marks the sheets with words from John 19:15: "Crucify him, take him away, crucify, crucify him, crucify, take him away and crucify him." In his use of paper for the juridical proclamations of Christ's crucifixion, Mantegna recognizes it as a support whose antique roots resonated with the manuscripts written in the languages of the Holy Land on paper, and presented as gifts to his patrons. The unfolded sheets of paper framing Christ also reverberate with the *gride,* a Gonzaga tradition of writing new laws on paper and pasting the proclamations in the city, where they would be shouted on Saturdays for those who could not read.[40]

Mantegna also continues the tradition of Peter the Venerable, in which paper is explicitly connected to Jewish texts and law. The paper crowns worn by Christ's handlers are inscribed with a pseudo-script that Nagel suggests is an evocation of Syriac script also known as Aramaic. The specific association of Mantegna's pseudo-script with Aramaic and not Hebrew indicates the predominantly Ashkenazi Jewish population in Mantua, who were associated with the Talmud, written in Aramaic.[41] In turn, the paper crowns evoke the famous trial of Jan Hus (1369–1415), who was executed for heresy at the Council of Constance. In the accounts of his execution, the eyewitnesses describe a paper crown marking him as a Judas and traitor.[42]

As one fifteenth-century text remarked, the paper crown was "light and easy," as opposed to the "heavy and oppressive crown" worn by Christ on his way to Calvary.[43] The importance of Hus's paper crown became an iconographic motif in illustrations of his trial and execution, and the paper crowns of Christ's accusers in Mantegna's painting resonate with this imagery that circulated in both manuscript and print (figs. 51, 52).[44] They signify Judas and heresy as the "lightness of paper," and its inherent mobility becomes a contrast to the weight of Christ's crown of thorns, cross, and Passion. Although paper

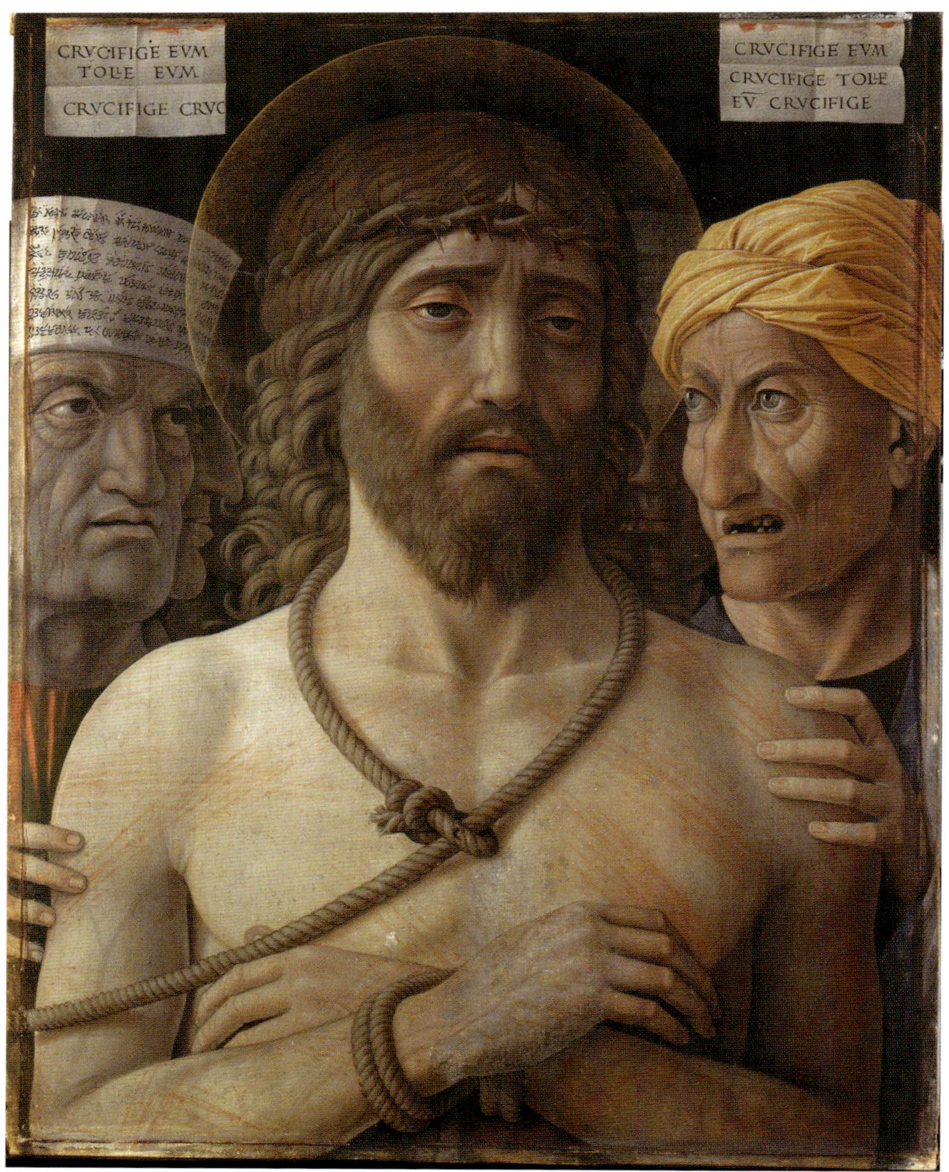

Figure 50. Andrea Mantegna, *Ecce Homo*, ca. 1500. Oil on canvas, 21¼ × 16 9/16 in. (54 × 42 cm). Musée Jacquemart-André, Paris.

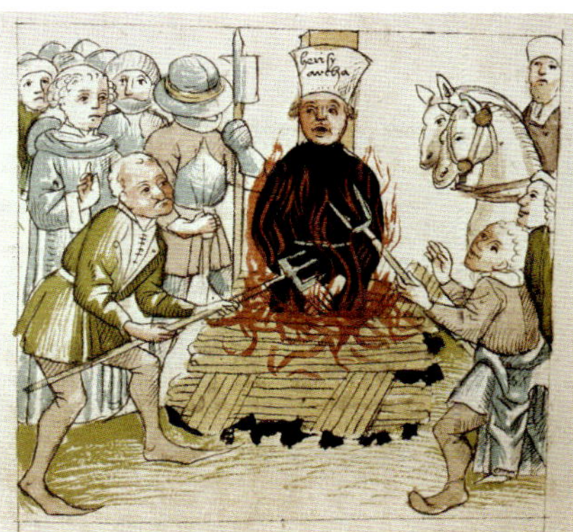

Figure 51. Ulrich von Richental, Fol. 58r: "The Burning of Jan Hus at the Council of Constance in 1415," in *The Chronicle of the Council of Constance by Ulrich von Richental* (Hs.1), ca. 1464. Watercolor and ink on paper. Rosgartenmuseum, Konstanz, Ger.

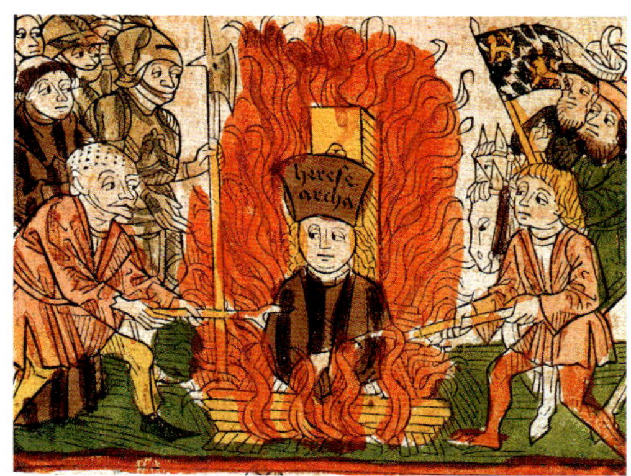

Figure 52. Ulrich von Richental and Anton Sorg (printer), "The Execution of Jan Hus," in *Das Concilium zu Konstanz,* ca. 15th century. Woodcut and watercolor on paper. Biblioteca Nazionale Braidense, Milan.

would transform diplomacy and correspondence across Europe, it maintained its traces to Judaism and a history of the book that was widely denigrated as false prophecy.

Yet paper was active as a diplomatic agent in the daily life of the court. In the Camera degli Sposi (1465–74), Mantegna positions the Gonzaga family's power through their paper diplomacy. The portraits depicting the Gonzaga family span across multiple walls in the chamber of Ludovico Gonzaga (1412–1478), united by the narrative device of the letter. To the left of the entrance, the family gathers as Ludovico receives a letter from his secretary, setting in motion a physical correspondence that plays out across the room (fig. 53). In the adjoining wall, Mantegna inserted a small letter into the hands of Cardinal Gonzaga with the inscription "*A[nd]rea me pi[nxit]*" (Andrea painted me) (fig. 54). Through the painted depictions of letters, Mantegna conveyed the significant role of the Gonzagas in European diplomacy. As Daniela Ferrari maintains, like many *seigneuries,* the establishment of paper archives was central to the Gonzagas' legitimization of their rule.[45] European courts manifested their power through an ability to control and archive paper networks.

Yet paper coexisted in heterogeneous worlds in the fifteenth century, from the antiquity of an Arabic version of Luke to the epistolary networks of the courts. Although paper archives would establish order and place, the complexity of paper in the early Renaissance undermined this projection of order. In the work of Simone and Mantegna, their acute awareness around paper's origins was instigated by a media revolution. For Simone, it was a scriptural transformation brought on by the rise of the notary and the introduction of paper mills to Italy. For Mantegna, it was the upheaval introduced by print and the emergent dominance of paper over parchment. Both artists worked during transitional periods, which made them acutely attentive to paper, and the ways in which it not only transmitted history but also itself imparted meaning.

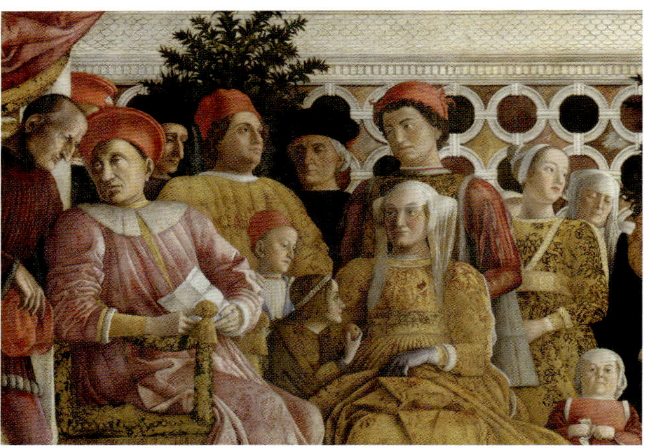

Figure 53. Andrea Mantegna, "Ludovico Gonzaga Receives a Letter," in the Camera degli Sposi, 1467–74. Fresco. Palazzo Ducale, Mantua, Italy.

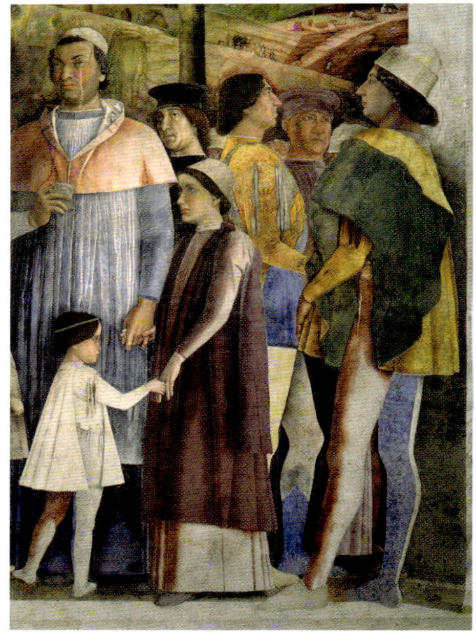

Figure 54. Andrea Mantegna, "Cardinal Francesco Gonzaga Holds a Letter with the Words '*Andrea me pinxit,*'" in the Camera degli Sposi, 1467–74. Fresco. Palazzo Ducale, Mantua, Italy.

CHAPTER III

The Model of Loss in Late-Medieval Drawing

In one of the first appearances of the word "paper" (*papier*) in the French language, a poet calls on his secretary: "Find some paper; I wish to write." This command is uttered in *Le livre dou voir dit* (ca. 1361–65; *The Book of the True Poem*) by Guillaume de Machaut (ca. 1300–1377), a work that contains the poetry and epistolary correspondence exchanged between an aging poet and his younger, admiring lover. In the instant when the poet calls for paper, he is agonizing over being apart from his beloved. As he tells his secretary: "I cannot keep silent. I believe I am forgotten by the beautiful woman I never forget. Find some paper; I wish to write."[1] In his professed anxiety over their physical distance, the poet relies on word and image as a means to salvage the imminent possibility of forgetting. With the exchange of letters and small painted portraits, the poet and his lover send one another constant physical reminders of their presence.[2] When he calls for paper and ink, Machaut's narrator takes in hand materials with which he may act out against the threat of being forgotten by his beloved.

This anxiety around neglect and forgetting, which Machaut plays out in regards to his paramour, is also at the heart of his own legacy, as he was devoted to overseeing the physical construction of his work into bound volumes. With a managerial concern considered indicative of an emerging sense of authorship, he directed the production of books that would contain his corpus for posterity. Unlike earlier compilations, Machaut's texts are not united by genre but by authorship.[3] And in his attention to the book as a physical object of transmission, Machaut recognized a connection between intellectual labor and its physicalization in manuscripts.[4]

Machaut's awareness surrounding the production and reception of his work continues in the writings of his successors, particularly Jean Froissart (ca. 1337–ca. 1405). Like Machaut, Froissart frequently refers to the materials of his craft as integral to writing. In *La prison amoureuse* (ca. 1372; *The Prison of Love*) the narrator states: "I didn't delay at all / but took paper and ink (*encre*), and like / a ship at anchor (*ancre*) I stayed put / until I had written / just what you see here."[5] Froissart plays with the timbre between *encre* (ink) and *ancre* (anchor), so that the narrator's task of writing secures the liquid medium of ink on the paper. Froissart fixes fluid, amorphous thought expressed by ink into contoured lines. He also draws a direct line between the moment of the work's writing—the taking of paper and ink—and the temporality of the viewer's perception, seeing "just what you see here."

Paper, the material that produced a revolution in written correspondence in fourteenth-century Europe, plays a pivotal role in the work of both Machaut and Froissart, as it often appears in moments of haste and desire. Writing on paper becomes defined against the careful copying and binding of texts on parchment. Froissart's narrator often transcribes his words first written on paper onto fine parchment, enclosing them in elaborate manners. In one example, the narrator mentions that once he was satisfied with a *lai* (a lyric narrative poem), he had it "copied onto fine parchment, / then tied up with silk ribbon."[6] Froissart frequently incorporates into the text details about the physical construction of his work, so that the process of writing and the final product are inseparable. In a humorous narrative in which a disfigured florin recounts all of the poet's expenditures, the coin describes the making of books as the most worthy expense: "First of all you have made books / Which cost at least seven hundred pounds. / There you spent your money well; / I admire this above all else, / For you made out of them many stories / From which there will still be memory / Of you in days to come / And you will make people remember."[7] As his account makes clear, the monetary cost involved in book production is offset by the promise of remembrance. And in the opening lines to his lyric *Le joli buisson de jonece* (1373; *The Bush of Youth*), Froissart once again intertwines the material of writing and memory: "I remember the experiences / Of times past. Thus it is fitting for me, / While I have a sound mind and memory, / Ink and paper and a writing desk, / A pen knife and a well-trimmed pen."[8] The tools for Froissart are not only his intellect but also paper and ink, penknife and pen: physical objects that become the writer's weapons against oblivion.

Machaut's and Froissart's acute awareness of the resources of their craft was shaped by their work as *clerc-écrivains,* or clerk-writers. Both men held roles in which they were not only literary figures but also administrators. Machaut's attention to the storage of his texts may be contextualized within his early professional life spent as a clerk for the king of Bohemia, John of Luxembourg (1296–1346), for which the successful transmission of texts was central to his job.[9] In turn, Froissart was a *clerc de la chambre,* or secretary, to Philippa of Hainault, wife of Edward III. The materials of paper, parchment, and ink were not only the materials of the writer but also the record-keeper. It was this dual valence of ink fixed on paper as both a source of literary invention and a documented record for posterity that made Machaut and Froissart acutely attuned to their craft. The preservation of documents was as vital as the intellectual appreciation of form. Clerk-writers such as Machaut and Froissart made "books, not performances."[10] Froissart, for example, explicitly expressed concern with his works' transmission to future generations.[11] At the end of *La prison amoureuse,* Froissart writes: "I would be most unhappy if, after having put my time, my heart, my love, and a whole period of my life into composing and writing *The Prison of Love,* both the prose and the poetry, in honor of my lady and Rose, no one were ever to see the fruits of my labor."[12]

While artistic process and memory are at the heart of fourteenth-century French literature, it has not been considered how the simultaneous emergence of drawing as a medium devoted to fragmentary experience captured in sketchbooks also engages with concurrent questions of inscription and remembrance.[13] This was a moment when the act of drawing became the object of drawing. While the material composition of writing was pivotal to the development of French literature circa 1400, a corresponding attention emerges in the work of artists collected and appreciated by the same courtly clientele, such as the Duke of Berry and his entourage.[14]

Machaut and Froissart attended to the role of manuscript production in transmitting their works to posterity because they recognized that authorship was tied to the physical object of the book.[15] This physicalization of their work indicates a shift from the primarily performative tradition of the courtly *troubadour* to the written work of the *rhétoriqueur.*[16] Whereas the poet's voice was integral to the lyrics of the troubadour, Machaut and Froissart worked against this ephemeral performance by compiling their work into books.[17] While the book also is a space of performance, it is not relegated to the temporality (or geography) of the human body. Books bear their own time and can travel apart from their authors. The performance of parchment and ink in the absence of the author's body guides the structure of Machaut's and Froissart's work, which is always constructed around the exchange of letters, rondels, and *dits* between poets and lovers, poets, and patrons. Like the written missals that form the edifice of Machaut's and Froissart's writings, their volumes would make the authors present, despite their physical absence. The consideration of transmitting the written work to posterity not through the oral culture of the troubadour lyric but instead through the material of paper and parchment, ink, and binding offers a parallel by which to examine late-medieval draftsmanship.

The Performance of Line

Medieval histories of drawing are complicated by an inability to define drawing.[18] For the most part, designs on parchment by illuminators remained unseen, as they were covered with paint and gold. As Eberhard König provocatively asks: "What kind of work should we call a drawing? Is it any initial stage of a work to be finished with layers of paint, not intended to exist on its own, but left to its particular status just by chance? Or is it any relic from a painter's process, such as a casual study of some detail, a preliminary sketch for a composition, or a *vidimus* that was part of the contract between workshop and patron?"[19] The problem is not only the lack of a definition but also the dearth of drawings that survive, which leads to the supposition that drawings were not valued. Yet the absence of drawings is not a measure of their worth. Instead it suggests that they demonstrated qualities that do not accord to contemporary concerns with binding, preservation, and conservation. As Maryan Ainsworth notes, "Netherlandish artists seem to have used the grounded panel as their sketchpads, working up drawings there, as much as on paper."[20] Painting over drawings does not necessarily indicate that they were not appreciated, but rather that they were ephemeral. Like the oral performance of the troubadour, drawing was a transitory act that would always be delimited by the necessary presence of the draftsman's body, a demonstration of memory and imagination honed through years of labor and work. Images were contained within the intellects of working craftsmen, who inscribed onto parchment and other surfaces the models in the mind, only to have these drawings erased or covered by paint.

Yet the importance of this culture of erasure has not been considered as a defining feature of late-medieval drawing. Instead, art historians focus on model books as sites for preserving and transmitting, presupposing that there *must* have been model books to transfer designs from "manuscript to manuscript, from atelier to atelier, from place to place, and from generation."[21] In marked contrast to this supposition, Ludovico Geymonat argues that model books do not survive because draftsmen relied on their memories and imaginations rather than on collected drawings.[22] The artists themselves "were the real agents of the transmission of artistic forms and ideas throughout the medieval world."[23]

To support his argument, Geymonat examined the *Wolfenbüttel Musterbuch*, which contains twelve parchment pages covered with about thirty figures documenting the peregrinations of an itinerant artist (perhaps Saxon, perhaps Italian?) through Venice and the Balkans (fig. 55).[24] The drawings exist today because the parchment was reused by a scribe, who copied the text of Guillaume de Saint-Thierry's *Epistola ad fratres de Monte Dei* (1144; *Letter to the Brothers of Mont-Dieu*) directly over the figures. The parchment quire was then bound into another manuscript, which allowed for the survival of both the drawing and the text.[25] Geymonat argues that the *Wolfenbüttel Musterbuch* does not demonstrate the existence of medieval model books, but rather drawing as a process that allowed artists to inscribe forms into their memories and imaginations.[26] The survival of the drawings is contingent. They exist today only because the parchment was repurposed for a written text, and like Pliny's *Natural History*, the drawings entered into the palimpsestic culture of the late-medieval scriptorium. The physical survival of the drawings was not important, because the artist's body and mind were the model book. Just as the troubadour's body and voice were the vehicle that realized the performance of late-medieval lyricism, so the draftsman's body and hand were the vehicles that performed the trace of line.

Yet just as Machaut's and Froissart's work illustrates a movement from the oral performance of the poet to the written work of the clerc-écrivain, so draftsmen in the same period demonstrate an acute awareness about containing, binding, and transmitting their designs to successors (beyond the circulation of the workshop patron). This engagement with a shift in drawing practice is revealed in three surviving boxwood drawing books—in Berlin, New York, and Vienna. These books bridge two distinct periods of the artist's workshop. The first period encompassed drawing on surfaces such as panel, wax, and wood, materials that made a process of drawing defined by erasure. The second period was realized in drawing on paper, which allowed an increase in the conservation of sketches, drafts, and drawings for posterity. The preservation of the boxwood books (an erasable surface) demonstrates an emergent interest in keeping the physical traces of the artist's labor for posterity, in building an archive of the artist's workshop. Yet this conservation of boxwood only emerged in the context of the paper archive.

Auctors and the Scene of Writing

The opening image to a boxwood drawing book in Berlin depicts two evangelists at work in a scriptorium (fig. 56).[27] One sits with his back to the viewer,

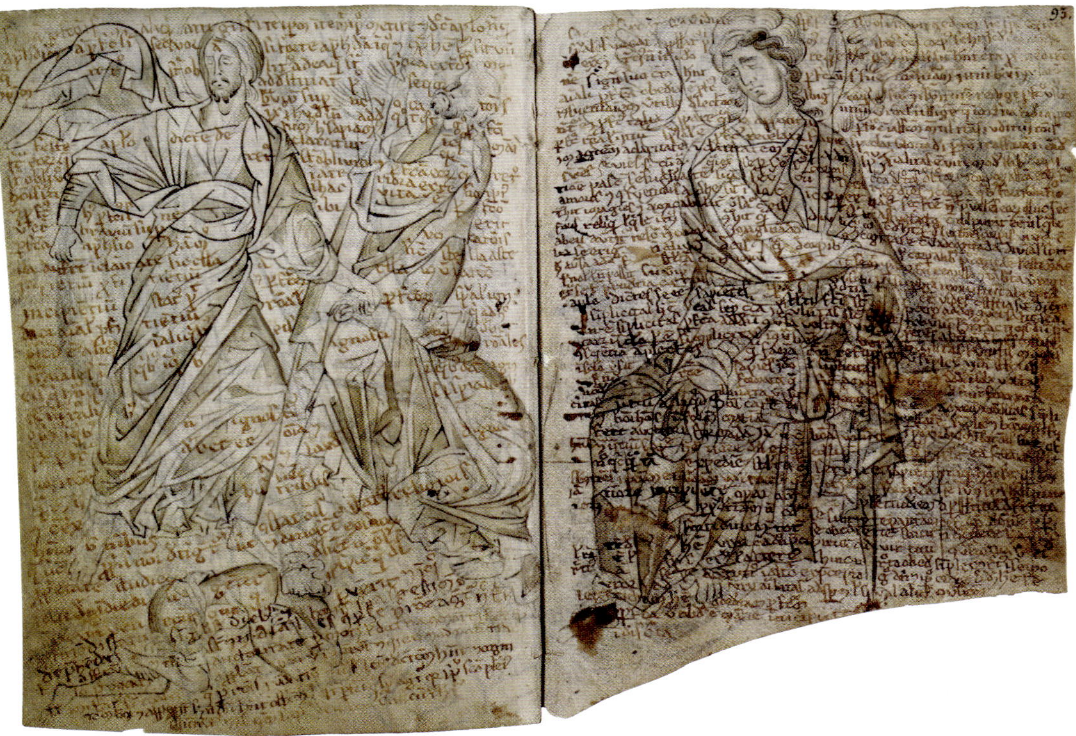

Figure 55. Fol. 92v and 93r in *Wolfenbüttel Musterbuch* (Cod. Guelf. 61.2 Aug. 8°), ca. 13th century. Ink on parchment. Herzog August Bibliothek, Wolfenbüttel, Ger.

bent over his parchment scroll. The other hunches over his desk, in profile to the viewer, a sense of fatigue reflected in his elongated face. Yet these are not mere scribes copying texts. As the inscriptions around their haloes denote, these are portraits of the evangelists Matthew and Mark. The angel dictates to Matthew while he writes, a mode of transmission that reflects the scribes who copied texts read aloud by abbots or other administrators in the scriptorium.[28] With the weariness delineated across Matthew's lean face and his stooped posture, the artist stressed the tedium of transcription.[29] Yet the boxwood folio also considers Matthew and Mark as authors, a role highlighted by the names inscribed in their haloes.[30] In turn, this naming structures the draftsman's own relationship to the image as he inscribed his own—Jaques Daliwe—on the lower border.[31] As the first scene in the boxwood book, the artist situates the composition within the world of the scribes. Unlike later artists who established a lineage between their work and the evangelist Luke, who painted the Virgin Mary, Daliwe founds his authorship through the scriptorium. Although it bears marks of erasure and retouching, it is thought that a single artist worked on the folios over a period of about twenty years. On the one hand, the compositions within the book directly cite popular models associated with the Limbourg brothers. *The Flagellation* (fig. 57), for example, is modeled on an illumination from the Limbourg brothers' *Belles heures* (fig. 58). Yet the drawing book also displays original compositions, with a particular attention to facial typologies (fig. 59).[32]

While it could be dismissed as an idiosyncratic choice for Daliwe to initiate his sketchbook in the scriptorium, the importance of writing to contextualize draftsmanship emerges in the other surviving boxwood drawing book as well. This second surviving boxwood drawing book also opens with a scene of writing, and both books have been associated with workshops from the Duke of Berry's court (fig. 60). In the second book, in the Morgan Library and Museum in New York, the opening image of the *Virgin and the Writing Christ Child* has been attributed to Jacquemart de Hesdin (ca. 1355–ca. 1414), an illuminator cited in the inventories of the Duke of Berry.[33] Whereas the Berlin book begins with the evangelists in the scriptorium, the New York book commences with Christ as *auctor,* writing on a scroll.[34]

One scholar suggested that the Writing Christ Child reflected the thirteenth-century Franciscan theologian Bonaventure's description of Christ: "Only Christ is the teacher and the author."[35] Yet the relevance of the term "auctor" in this period extends beyond Bonaventure, as it was gaining new prominence in vernacular literature. Froissart was the first historian to designate himself as an auctor. In the prologue to *Chroniques* he wrote: "And so that future generations know who wrote this story and its author (auctor) was, I would like to introduce myself. I am Sir Jean Froissart, born in the county of Hainault, in the beautiful city of Valenciennes."[36] In his employment of "auctor" as a means to designate his labor, Froissart called on a term that traditionally had been reserved for writers from antiquity.[37]

Machaut and Froissart, like Bonaventure, discuss Christ as an auctor and a lyric poet. For Machaut, Christ is a lyricist who writes verses for the Virgin: "It is certain that Jesus Christ, as I find in my writings, says courteous and clear words about the lily in his little songs."[38] In her analysis of the dit, Sylvia Huot remarks that Christ becomes the "author and protagonist of historical writings and of exquisitely beautiful love poetry."[39] It is within this courtly context, in which the lover marks his progression in verse from the imperfection of earthly desire to devotion to the Virgin, that the emergence of the Writing Christ Child as a popular motif takes shape.[40]

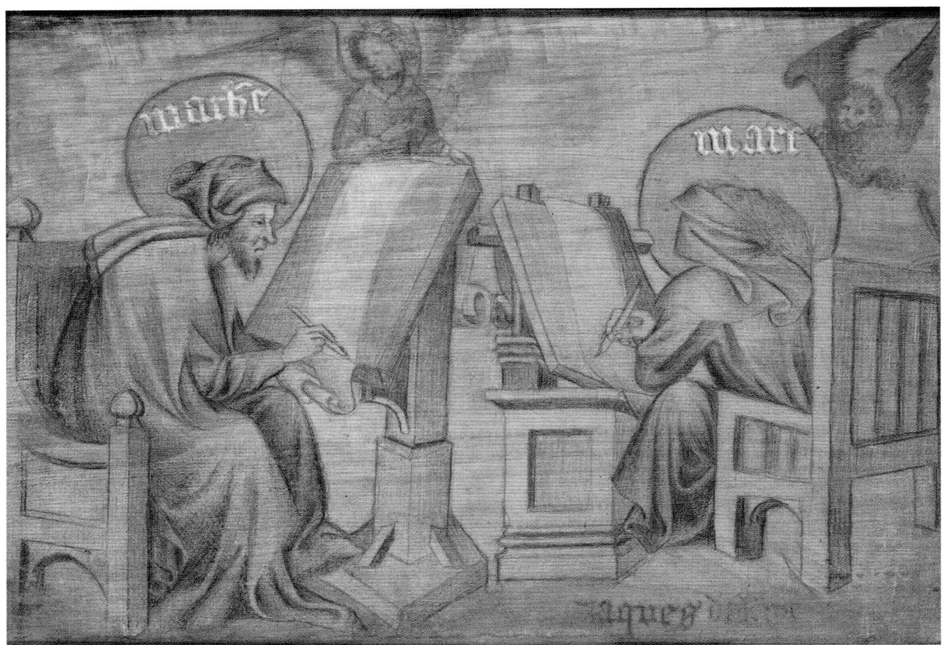

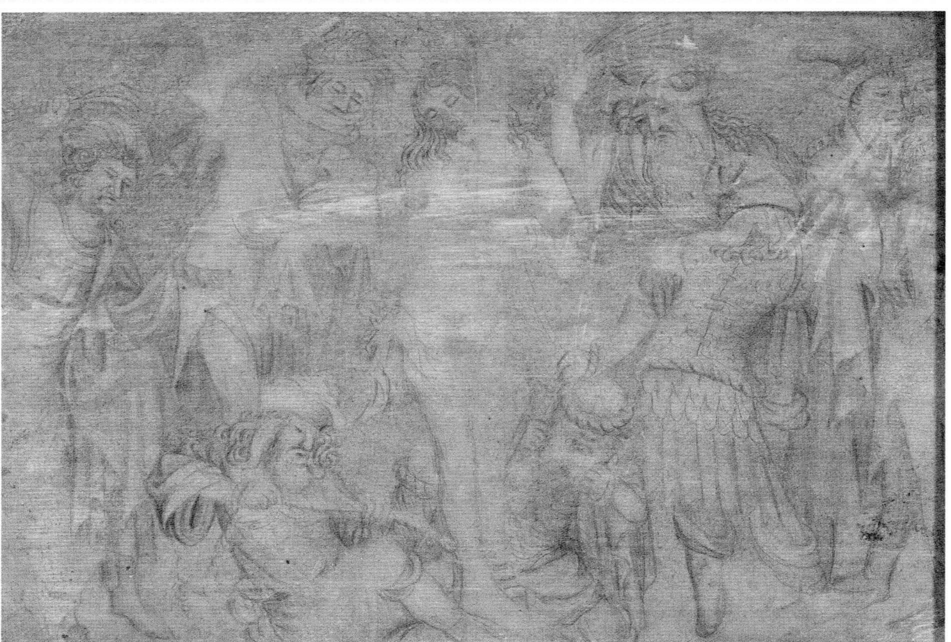

Figures 56 and 57. Jaques Daliwe (?) in the *Berlin Sketchbook*, ca. 1400–20. Metalpoint on boxwood. Staatsbibliothek zu Berlin, Prussian Cultural Heritage Foundation, Manuscript Department, Libr. pict. A 74. Top: Fol. 1B, "The Scriptorium"; bottom: Fol. 2a, "The Flagellation."

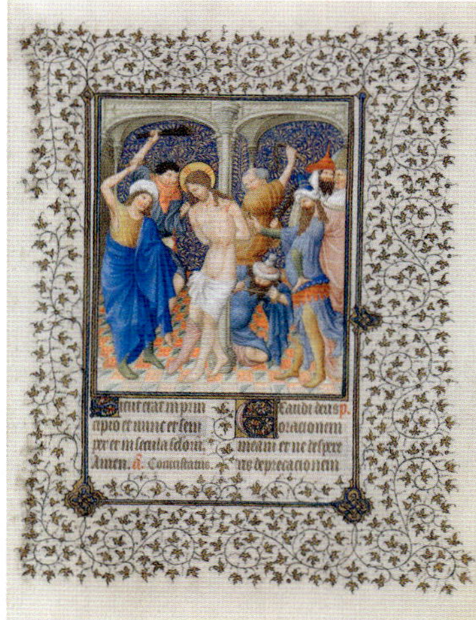

Figure 58. Limbourg brothers, Fol. 132: "The Flagellation," in *The Belles heures of Jean de France*, duc de Berry, 1405–8/9. Tempera, gold, and ink on vellum. Cloisters Museum, New York.

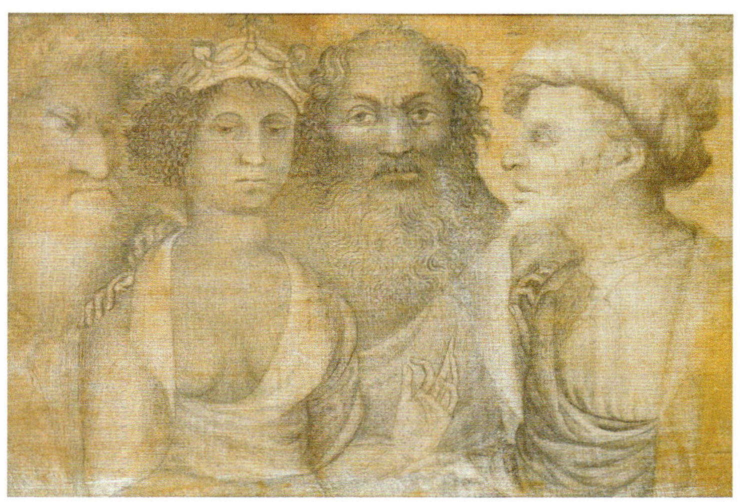

Figure 59. Jaques Daliwe (?), Fol. 9a: "Study of Facial Typologies," in the *Berlin Sketchbook*, ca. 1400–20. Metalpoint on boxwood. Staatsbibliothek zu Berlin, Prussian Cultural Heritage Foundation, Manuscript Department, Libr. pict. A 74.

Unlike the Berlin boxwood drawing book, the New York book consists of multiple authors. In its display of different hands, the boxwood book charts one of the most important stylistic developments in late-medieval French manuscript illumination from Jean Pucelle (ca. 1300–1355) to Jacquemart de Hesdin—from the stylized curves of Pucelle's grisaille to the Bohemian faces of Jacquemart, with their marked, individualized expressions.[41] The first hand identified in the book resonates with Pucelle's curvaceous and supple line, as soft drapery folds and a strong linear outline play against the masterful grisaille modeling that came to define his work. The subjects of this hand are chivalric scenes of knights, courtship, and ornamental motifs. The foliated leaf ornament (fig. 61) and the jousting (fig. 62) present *exempla* that could be reinserted and recomposed into a variety of compositions.

The second hand has been identified as that of Jacquemart. As opposed to the earlier drawings, which are abstractions of drapery, ornament, and figures, the later hand displays an acute interest in facial studies. Whereas the earlier scenes are generalized moments of chivalric courtship, the later ones respond to events at the court of Charles VI. The coexistence of both distinct styles and

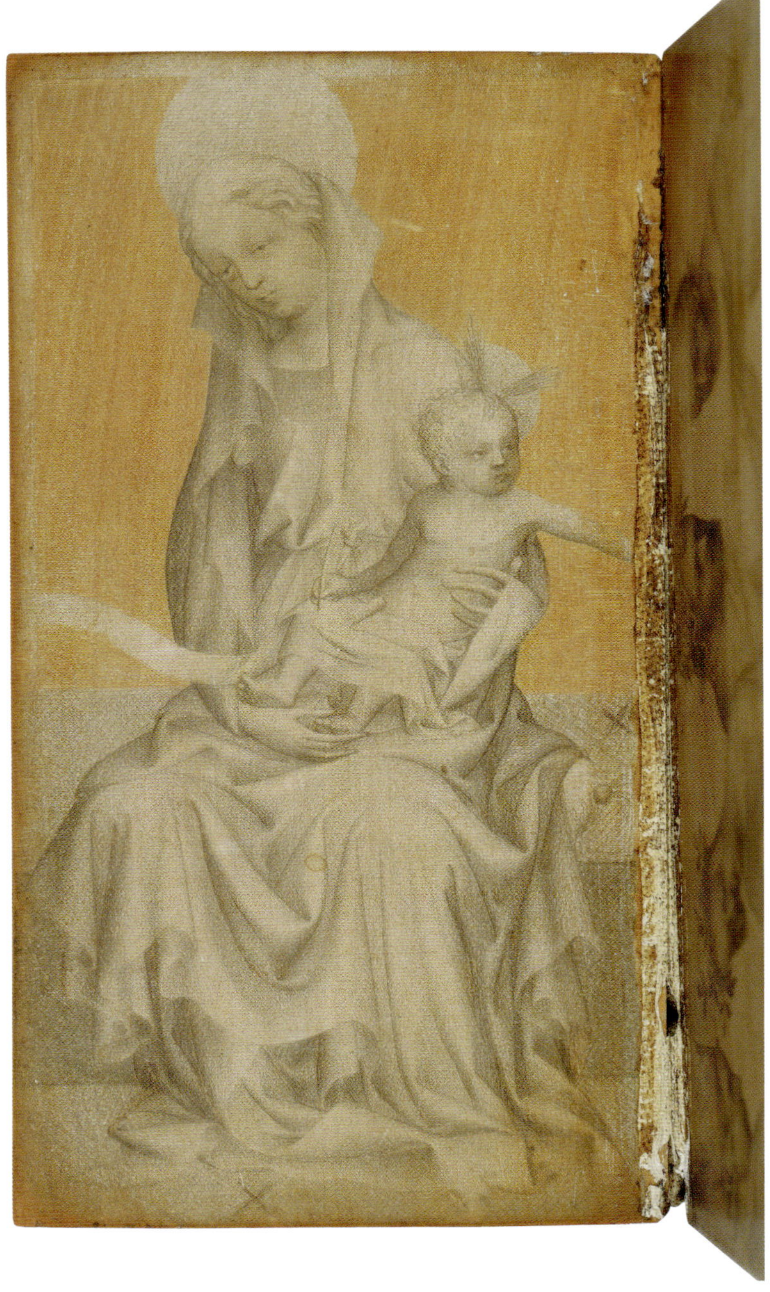

Figure 60. Jacquemart de Hesdin (?), MS M.346 fol. 1v: "Virgin and the Writing Christ Child," in the *Boxwood Sketchbook*, ca. 1400. Metalpoint on prepared boxwood. Morgan Library and Museum, New York.

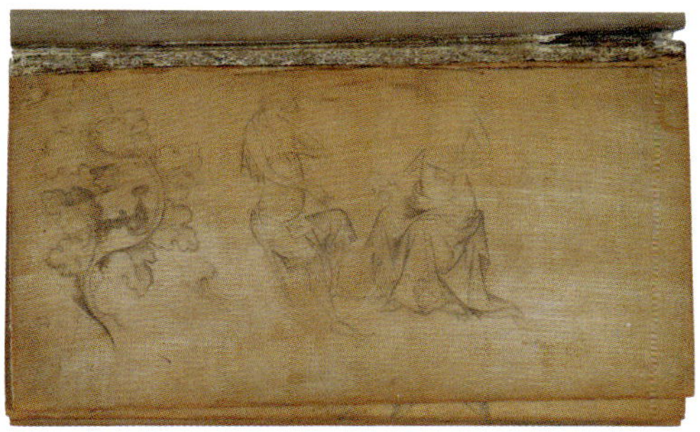

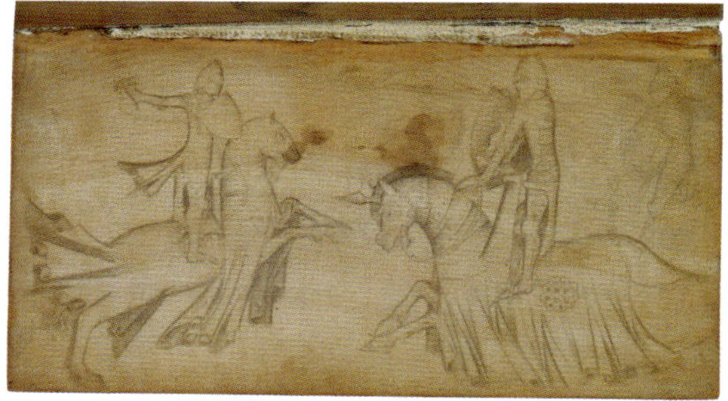

Figures 61 and 62. Anonymous, MS M.346 in the *Boxwood Sketchbook*, ca. 1400. Metalpoint on prepared boxwood. Morgan Library and Museum, New York. Left: fol. 5v, "Foliate Ornament and Seated Figures"; right: fol. 5r, "Jousting."

subjects within a single work demonstrates how Jacquemart established his own model of authority through a direct engagement with the work of his predecessors. As Millard Meiss observed, Pucelle initiated a "new art (*ars nova*)" that culminated in Jacquemart, as he became newly invested in the "rhetoric of painting."[42] Whereas the Pucelle-style works belong to a lineage of forms that scholars have called exempla, the Jacquemart drawings evoke vernacular histories from the court of Charles VI recorded in Froissart's *Chroniques,* the epic narrative detailing the first part of the Hundred Years' War. The boxwood drawing book therefore not only reveals a stylistic shift from Pucelle to Jacquemart, but also a major change from the exempla to a historical awareness of drawing's role in capturing history.[43]

The image that directly cites Froissart's *Chroniques* is folio 3v (fig. 63), which depicts a savage ball, a tradition Froissart memorialized. Composed of wild men crowned with leaves, the drawing illustrates the dance, in which courtiers dressed in linen suits covered with flax in imitation of hair. The festive occasion was marked by tragedy, as the revelers danced into the main hall only to be lit on fire as the curious Duke of Orléans came closer with his torch to see the masqueraders. Four of the masked men died in the tragedy.[44] Whether the drawing was made before or after the tragedy, it persisted in its wake.

Yet this is not the only drawing that speaks to events at the court of Charles VI. Michael Camille and others have suggested that folio 2r depicts the meeting between Charles VI and his young bride, Isabeau (fig. 64).[45] Camille directly relates the drawing to Froissart's description, in which the "King went towards her and, taking her by the hand, raised her up and looked at her long and hard.

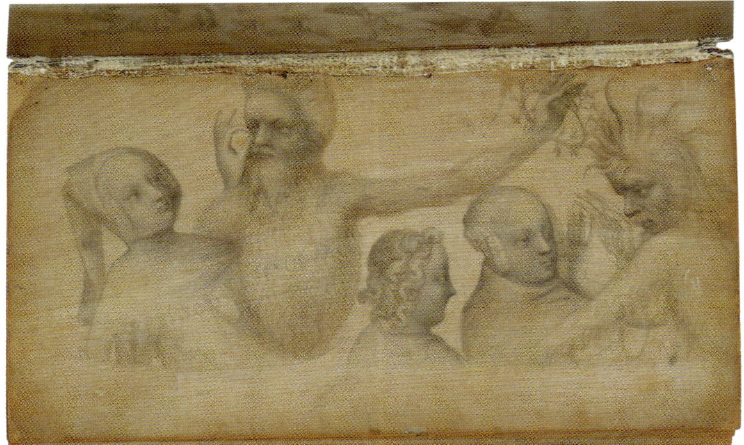

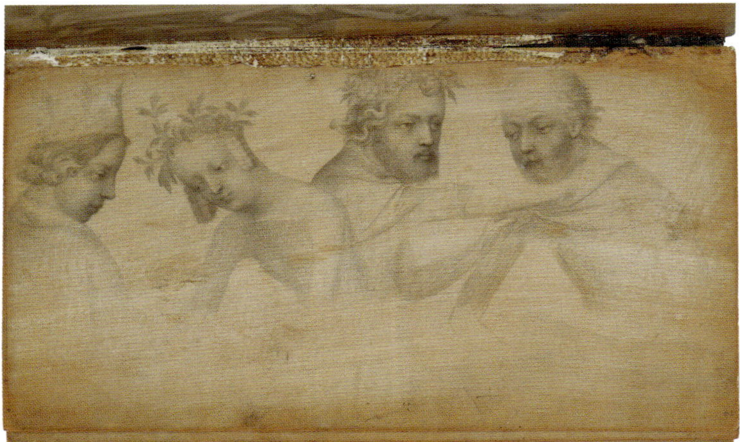

Figures 63 and 64. Jacquemart de Hesdin (?), MS M.346 in the *Boxwood Sketchbook*, ca. 1400. Metalpoint on prepared boxwood. Morgan Library and Museum, New York. Left: fol. 3v, "Savage Ball"; below left: fol. 2r, "Charles and His Bride Isabeau (?)."

With that look love and delight entered his heart. He saw that she was young and beautiful and was filled with a great desire to see her and have her."[46] Camille also tentatively suggested that the other crowned male figure was the brother of Charles, the Duke of Orléans, who was rumored to have had an affair with the queen. In drawing the boxwood book into the salacious world of court gossip, Camille notes that the work "was surely not meant to be made public, but was kept in a closed box of precious drawings that artists such as Jacquemart increasingly guarded as their private property."[47]

As Froissart wrote his monumental text on the diplomatic events of the Hundred Years' War, a draftsman consigned into a small and obscure boxwood book a collection of sketches that parallel the descriptions of events in the *Chroniques*. The drawings do not copy, reiterate, or contain models for the workshop. They do not offer iconographic formulas for mythological and biblical histories. They were never reused for later illustrated editions of the *Chroniques*. Instead they evoke events from the king's court that would have resonated with a circle of cognoscenti. As these authors and draftsmen moved into the particular world of contemporary history, they at once depended upon previous models (Machaut, Pucelle) by which to inscribe their work into the present, while also marking it as something new, as auctors of history.

Erasure defines both the Berlin and New York drawing books. Ultraviolet images of the panels have revealed traces of earlier drawings on many of the leaves, illustrating that the second set of images was drawn over previous designs.[48] Both boxwood books were shaved down at certain moments in order to reuse the material. Each draftsman made this process integral to the drawing book, keeping both stylistically archaic and novel formulations within his work, making clear the selective process of memory. These books act as sites of erasure and containment, at once ephemeral and memorial, marking artistic practice as defined both by destruction and survival.

In antiquity, the term "*buxum*" (boxwood) often referred to a wax tablet.[49] Until the invention of paper, the wax tablet was the primary site of composition, whether for a document, letter, or poem.[50] The act of drawing on wax embodies draftsmanship as a metaphor for creation, as evoked by Alain de Lille (ca. 1130–1202/3), who describes nature's creative process as an infinite method of realization and disappearance: "With the aid of a reed-pen, the maiden called up various images by drawing on slate tablets. The picture, however, did not cling closely to the underlying material but, quickly fading and disappearing, left no trace of the impression behind. Although the maiden, by repeatedly calling these up, gave them a continuity of existence, yet the images in her projected picture failed to endure."[51] This portrayal of drawing on erasable tablets resonates with the practice captured within the Berlin and New York boxwood books, as the draftsman planed down the boxwood to sketch over previous compositions, a course of coming into being and eradication. Yet the dominance of the wax and wooden tablet declined as the paper industry in Europe developed. This shift from erasable tablets to paper is documented in the royal accounts of France. The last mention of purchasing wax tablets was in 1359, when there is a record of two tablets and two metal styluses acquired for the king. In 1380 Charles VI's clerk buys paper and parchment but no wax tablets.[52] In turn, as paper supersedes the tablet in the administration of the court, drawings become a body of work to be kept, collated, named, dated, bound, and preserved.

Leather Coffers

The third surviving boxwood drawing book, in the collection of the Kunsthistorisches Museum in Vienna, is remarkable as it is composed of both boxwood and green prepared paper and still has its original leather coffer (fig. 65).[53] In its combination of tablet and paper, this book bridges two traditions, looking back toward the wax tablet and boxwood while also modernizing it with green prepared paper, which Cennino Cennini described as the "gateway of painting."[54]

Boxwood is the surface that Cennini recommends for beginning draftsmen, as its durability allowed practice on an economical material that could be shaved down, and so consistently reused. Cennini's adherence to boxwood as a surface for composition extended beyond the artisan's workshop in the fourteenth century to the practice of antique poets and rhetoricians. In perhaps one of the most prominent discussions from antiquity of boxwood's *materia,* Propertius mourns the loss of his boxwood tablets: "As me, some miser is writing his accounts on them and places them among his unfeeling ledgers! If anyone returns them to me, he will be rewarded with gold: who would wish to keep wood instead of wealth?"[55] Propertius presents boxwood as an erasable surface employed by both the poet and the miser, inscribed both with the immortal words of the lover and the mundane numbers of the accountant. The poet acknowledges the relatively inexpensive value of the tablets, which attain their worth through Propertius's authorial inscriptions: "No setting of gold has made them precious: they were just cheap wax on common boxwood."[56] This history of writing on an economical surface, in which the words of the poet transcend the poverty of the materials, bridges both boxwood and paper. Yet the elegy over the loss of his tablets also reveals the distinction between wax tablets and paper, as Propertius expresses his concern over their erasure and reuse. As discussed, paper did not lend itself to this culture of the palimpsest. And in its employment of both boxwood and paper, the Vienna model book bridges two traditions, embedding within its tablet wood frames for the modern surface of paper.

As Cennini notes, "Moving step by step toward enlightenment, so that you begin to want to discover the basis and the gateway of painting, you need to take up a different way of drawing from that which we have been discussing up

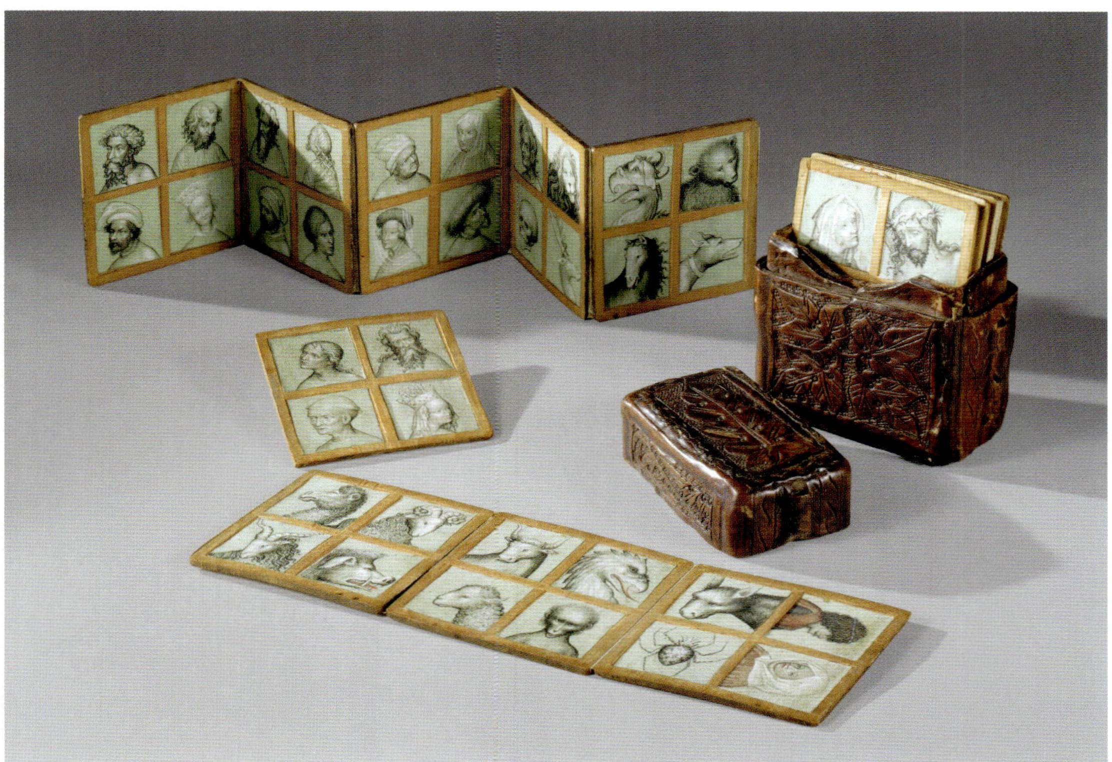

Figure 65. Vienna model book with its leather box (Kunstkammer, 5003), ca. 1400. Metalpoint on green prepared paper, mounted on boxwood panels and kept in a stamped leather case. Kunsthistorisches Museum, Vienna.

to now, and this is called drawing on prepared paper."[57] The Vienna book, therefore, exemplifies the most fashionable mode of drawing circa 1400, working on paper prepared with a colored surface. Yet it also frames the tinted papers in the history of the wooden tablet.[58]

This drawing book is exceptional because its leather case was preserved with the drawings. While the coffers of the other two boxwood books no longer survive, Camille and other scholars suggest that they were kept in such containers. Indeed, the drawing book's delicate box is integral to the meaning of drawing itself. For the importance of the case as a means to demarcate precious objects was fundamental to poetry, and the evocation of these small leather chests in Froissart describes their importance for drawings as well as verse. Froissart writes in *Prison amoureuses:* "I put them [poems] in a little coffer which I had carefully made of smooth Danish leather (for I gladly keep myself busy with such things)."[59] Or as Froissart recollects: "I used to have a little chest, pretty and polished, made of boiled leather. It was long and narrow and I placed, as was my wont, all three ballads in it."[60] Making these leather cases was important to the poetic process for Froissart.[61] As Peter Ainsworth suggests, Froissart wanted his work "consigned, bound, preserved, enshrined—yet not quite *entombed* or *embalmed,* for it must also be transmitted to posterity."[62] According to Jacqueline Cerquiglini-Toulet, placing the poems in their leather case allowed them to be "preserved and amassed as capital. It has a practical purpose: to avoid dispersion and loss."[63] Just as the poem becomes an object for collection and survival, so too does the drawing.

The boxwood drawing books stand as models toward marking the authorial performance as an object. As paper became the defining material for archives and letters, it established itself in the workshop as a support for posterity. Drawing on paper replaced the previous tradition of the palimpsest conserved in the boxwood books, which intimate the history of drawing on erasable supports. The written culture of the clerc-écrivain and the production of manuscripts informs the structures of these boxwood drawing books, as the draftsmen called upon images of writing as a way by which to reconceive their role as artists. The boxwood drawing books stand on the cusp of a new form of drawing, in which the act is not a transient performance but a record to be kept. Artists no longer erased previous artists' work in order to conserve materials or to enact moments of displacement. Until the introduction of paper to the artist's workshop, drawing was an act embedded in loss. With paper, drawings became objects to be kept in their leather coffers, preserved—so as not to be forgotten.

CHAPTER IV

Albrecht Dürer and the Geography of Paper

During his first *Wanderjahre* in the early 1490s, Albrecht Dürer traveled to Colmar to meet Martin Schongauer (ca. 1445–1491), an innovative engraver whose fine lines and dense hatching influenced the young Dürer. But Schongauer died before Dürer arrived. Nevertheless, Dürer continually encountered the older artist through his surviving works on paper, an interaction that Dürer carefully marks. As he wrote on one drawing by Schongauer: "This was engraved (*gerissen*) by Handsome Martin in the year 1470, when he was a young journeyman. I, Albrecht Dürer, found that out and have noted it here in his honor in the year 1517."[1] Dürer inscribed two dates: the year when the drawing was made, and the year when he amended it, attesting to his knowledge of the work and establishing a relationship to Schongauer. Existing together on the single sheet, they signal the passage of time between its making and its reception, between the work of a young journeyman in the 1470s and the posthumous recognition of his work by an artist in 1517. The inscription also acknowledges a future viewer—artist, author, archivist—who will encounter the work after Dürer's death.[2] This drawing, however, no longer exists. But scholars know the inscription because it was recorded in inventories, and it demonstrates that the work became recognized not only in terms of the connection to Schongauer but also Dürer's caption.

Other drawings, however, do survive that also testify to Dürer's practice of marking works on paper with their artist and date. On a *Blessing Christ* by Schongauer, Dürer extolled: "This the handsome Martin made in 1469" (*Das hat hubsch martin gemacht jm* 1469 *vor*) (fig. 66). Similarly, in a whimsical delineation of a woman fanning a fire with a bird's wing, Dürer recorded Schongauer's monogram and the date 1469. And in Schongauer's copy after Rogier van der Weyden's *Christ in Judgment* from the Beaune Altarpiece, Dürer also appended Schongauer's monogram and the date 1469.[3] As Dürer's inscriptions reveal, Schongauer made these drawings as a journeyman while learning his craft. It is questionable whether Schongauer would have considered them as collectibles beyond the reuse of pattern and design in the workshop. Yet with his brief legend, Dürer removed the drawings from the circuit of designs to be employed by the workshop. Instead he made them illustrations of Schongauer's development as an artist. Moreover, he created a relationship between the stylistic development of Schongauer and himself.

Dürer likely obtained these drawings from the Michael Wolgemut workshop in Nuremberg, which in turn inherited them from the Hans Pleydenwurff workshop, where Schongauer worked between 1469 and 1470. It is perhaps through knowledge gained from the oral memory of the workshop that Dürer affirmatively dates the drawings to the year 1469 or 1470, a particular habit of dating that, as Christopher Wood points out, was unique to Dürer.[4] When Dürer archived the Schongauer drawings, he was not overtly interested in them as models for his own workshop. It is possible that he traced or copied the drawings, although no evidence of this remains, and he did not engage with them as he would with Mantegna's engravings. Instead, Dürer denoted the drawings as Schongauer's, gave them a date of execution, and filed them among his own drawings, illustrating a history of draftsmanship.

Paper provided a material that invited artists to keep other artists' drawings. Instead of erasing and drawing over Schongauer's designs—as happened on parchment, boxwood, or wax—Dürer dated and kept the paper works. This attention to drawing and its role in forming an artist's cannon also impacted Dürer's own approach to paper, and points toward the ways in which he framed his painted artistic commissions with his drawings. Yet Dürer's appreciation of paper extends beyond its convenience as a storage material for designs and compositions. For he was aware of the geographic origins of different types of paper technologies and the ways in which they informed stylistically distinct types of drawing, and, in turn, different concepts of artistic practice and labor.

Dürer's acute awareness of the geography of paper is most clearly demonstrated during his second trip to Venice (1505–7), when he adopted a new technology known as *carta azzurra,* or blue paper. He used this paper to make drawings for the three paintings that he executed in the Veneto.[5] The most

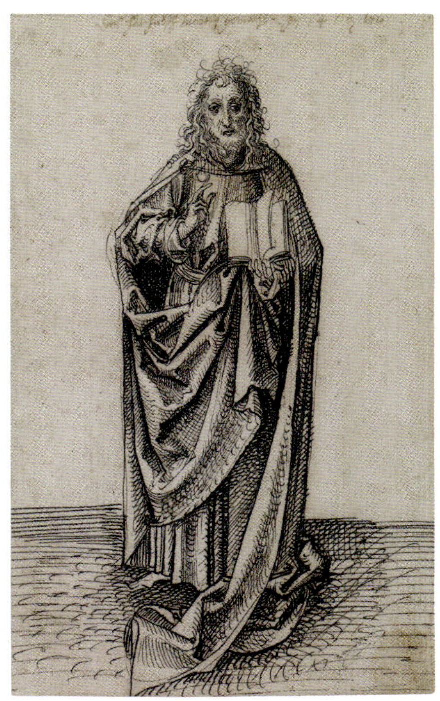

Figure 66. Martin Schongauer, Blessing Christ (with inscription by Albrecht Dürer). 1469. Pen and black ink, 8⅛ × 4⅞ in. (20.7 × 12.4 cm). British Museum, London.

prestigious of the commissions, the *Feast of the Rose Garlands* (fig. 67), was for the church of the German merchants in Venice, San Bartolomeo di Rialto.[6] The other two paintings are smaller in size, and their status as either commissions or gifts remains uncertain. In a letter to his friend Wilibald Pirckheimer, Dürer described one of the panels, the *Christ among the Doctors* (fig. 68), saying, "The like of which I've never done before."[7] In this composition, Dürer reveals a stylistic debt to Mantegna and Giovanni Bellini (ca. 1430–1516) with a horizontal format and close-up view of the figures. The third work is known as the *Madonna with the Siskin*, and also is rooted in Bellini-esque models of color and composition.

These three paintings completed in Venice are the only panels in Dürer's corpus for which he executed drawings on blue paper, a local Venetian commodity. Blue paper was central to papermaking in Europe since its arrival in the thirteenth century, and Venice was particularly renowned for its carta azzurra. It became well known as a drawing support in sixteenth-century Venice, as draftsmen studied modeling with the blue tonality, providing a middle tone between light and shade. Blue paper could also be cheaper to produce because

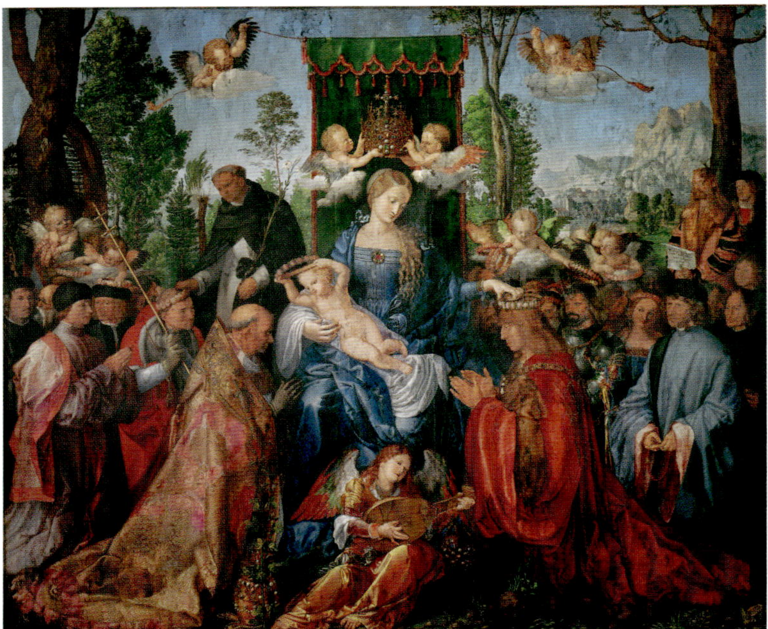

Figure 67. Albrecht Dürer, *Feast of the Rose Garlands*, 1506. Oil on wood, 63 13/16 × 75 5/8 in. (162 × 192 cm). National Gallery, Prague.

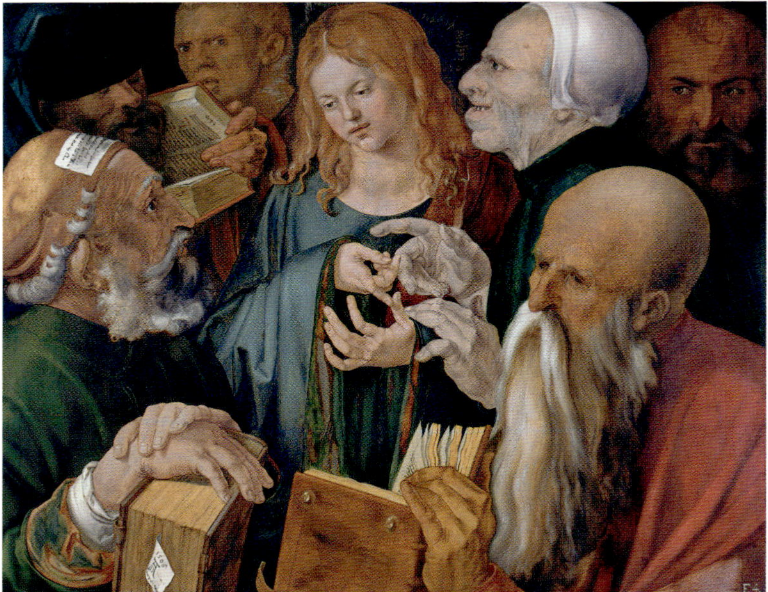

Figure 68. Albrecht Dürer, *Christ among the Doctors*, 1506. Oil on wood, 25 5/16 × 31 5/8 in. (64.3 × 80.3 cm). Museo Thyssen-Bornemisza, Madrid.

it did not require the careful selection of rags necessary in the production of expensive white paper.[8] Scholars also suggest that Venice was renowned for its blue paper because the advanced technology of fabric dyeing on the *terra firma* provided the foundation for colored paper.[9] Unlike the popular drawings on pink, brown, and reddish prepared grounds that dominated in Italy and Germany in the fifteenth century, the blue ground of carta azzurra is intrinsic to the paper, and not a separately applied pigment (figs. 69, 70).[10]

Although some scholars suggest that Dürer introduced drawing on carta azzurra in Venice, Catherine Whistler's scholarship on Venetian drawing establishes the prominence of blue paper in Venetian workshops by the end of the fifteenth century, particularly in the corpus of Vittore Carpaccio (1465–1520). In his drawings, Carpaccio exploited the mid-tone of the blue paper, against which he demarcated the bright-white highlights on the faces (fig. 71), while modulating areas of shadow and recession with black chalk and brush.[11] Yet where Carpaccio relied on line, Dürer pursued three-dimensional effects through chiaroscuro. Carpaccio's earlier drawings are linear; Dürer's are sculptural. In his drawing of Maximilian's hands from the *Feast of the Rose Garlands* (fig. 72), Dürer's white highlights break from hatching into painterly strokes, in which the grain of pigment and binder stands out from the paper's texture. Dürer exploited the plastic potential of the blue support so that the hands emerge from the page.

Although the carta azzurra drawings often are cited as some of the first surviving examples of preparatory drawings in Dürer's oeuvre, these works have little preparatory character.[12] Erwin Panofsky called them "equivalents, rather than as anticipations, of paintings."[13] And a close examination of the drawings' finish indicates that they were not produced for the paintings, but rather either in tandem with the panels or after their completion. They are not rapid compositional sketches, or iterative reconsiderations of form, but virtuoso performances demonstrating his full mastery of a new material. The drawings reveal a self-consciousness about their status as works on paper that both demarcates the creative process toward composing a painting, and also stands independent of the painted works as exemplars of draftsmanship.

One of the central features of the drawings is the carefully composed juxtapositions that Dürer created on a single sheet. Today, it is hard to imagine the original appearance of the drawing, as later collectors divided the sheets in two. But in his original compositions, Dürer consistently juxtaposed parts from his two major Venetian paintings on a single sheet of paper. In one study, Dürer contrasted Emperor Maximilian's hands from the *Feast of the Rose Garlands* to a hand holding a book from *Christ among the Doctors* (fig. 73).[14] On another sheet, Dürer contrasts the upturned gaze of the lute-playing angel

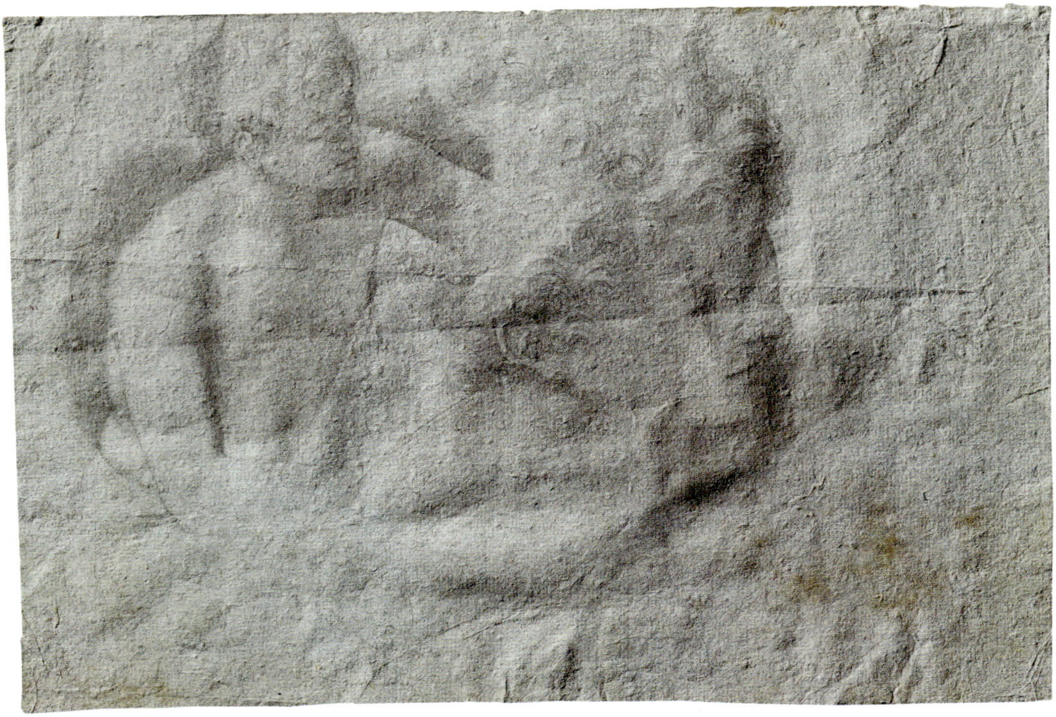

Figures 69 and 70. Cristoforo Roncalli, *Two Studies of a Youth*, ca. 1600–10. Black chalk on blue paper, 11¼ × 16⅛ in. (28.5 × 41 cm). Clark Art Institute, Williamstown, Mass. Above: image taken in raking light; Left: image taken in transmitted light.

from the *Feast of the Rose Garlands* (fig. 74) to the Christ Child's thoughtful downturned gaze in *Christ among the Doctors* (fig. 75). The diametrically opposed direction of their faces harmonizes across the sheet, as Dürer considers the realization of divinity with dimpled chins and slightly plump lips set into rounded faces (fig. 76).[15]

This care and attention to the harmonious placement of forms on a single sheet from two separate Venetian paintings suggests that Dürer did not make the drawings as studies. Instead they are paradigmatic vehicles of expression from his Venetian period. Further evidence of their lack of preparatory character is the state of the drawings, for they bear little trace of the wear and tear that might accompany preparatory drawings, such as the small punctures for pouncing or the light incision of a traced line. Workshop practices depended on techniques such as tracing and cutting, which destroyed many preparatory drawings.[16] Yet making *ricordi*, or drawings after finished paintings, was established in Venice (and elsewhere). Ricordi both generated a model for apprentices in the workshop and allowed the artist to reevaluate "his figural vocabulary."[17] The import of ricordi would have been even more vital for Dürer, as his workshop was based in Nuremberg and not in Venice. Making drawings after his paintings produced models for his workshop north of the Alps and allowed him to build an archive of the work he completed abroad. Just as Schongauer's drawings from his years as a journeyman in Nuremberg survived and testified to his artistic development, so Dürer's paintings produced abroad—archived by the artist himself—testified to the impact of Venetian artistic practice on his craft.

These drawings are a departure from Dürer's previous graphic works, which developed from his training as a goldsmith. Dürer's pre-Venetian drawings are bound to a vocabulary of precision and line that he abandoned in Venice in favor of coloristic effect. Three tones dominate the carta azzurra works: white, blue, and variations of gray. Yet they are studies of color and how to modulate light and shadow to achieve plasticity. Instead of depending on line and hatching, Dürer allows the middle tone of blue to create a balance against light and dark, examining the optical effect achieved through the surface of the paper. As Dürer demonstrates in his short excursus on color, he was primarily concerned with the movement of light and dark across the surfaces so as to "deceive the eye": "If you wish to paint in the sublime manner, such that is to deceive the eye, you must be very well instructed about colours . . . since in everything that creases and bends away from the eye there is light and dark. If that were not so, everything would look smooth and even, and in such a case one would not be able to recognise anything except merely how the colours are

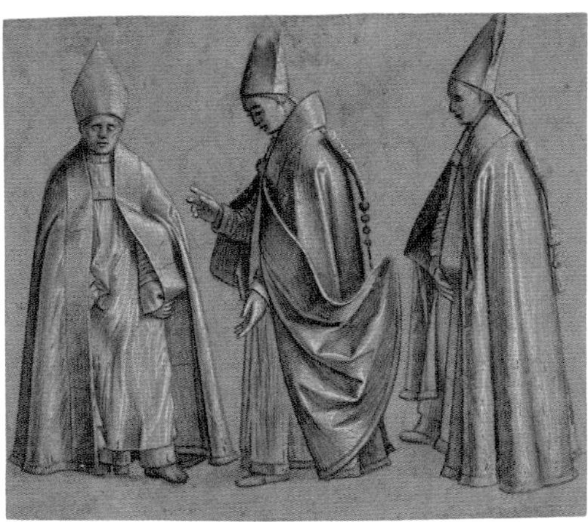

Figure 71. Vittore Carpaccio, *Three Studies of a Bishop*, ca. 1495–98. Black chalk on blue-gray paper, $7\frac{3}{4} \times 8\frac{5}{8}$ in. (19.7 × 21.9 cm). British Museum, London.

Figure 72. Albrecht Dürer, *Hands of Maximilian I from "Feast of the Rose Garlands,"* 1506. Brush drawing in gray and black with gray wash and white highlights on blue paper, $8\frac{3}{4} \times 10$ in. (22.3 × 25.4 cm). Albertina Museum, Vienna.

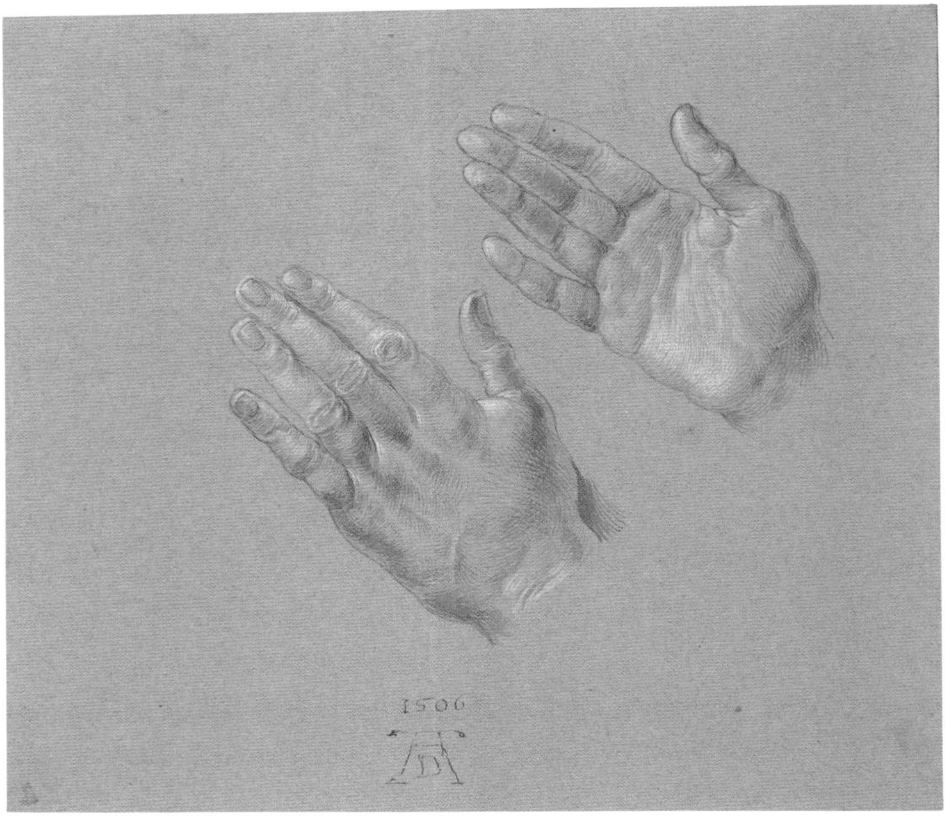

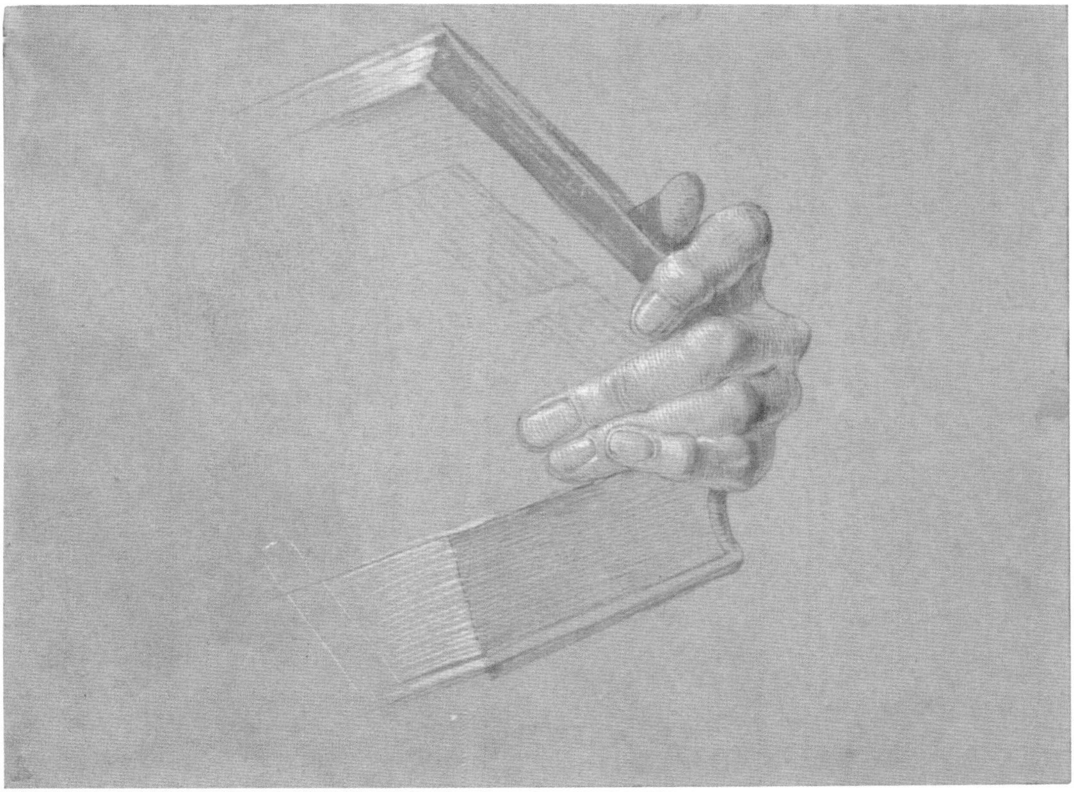

Figure 73. Albrecht Dürer, *Hands of Jesus from "Christ among the Doctors,"* 1506. Brush drawing in gray and black with gray wash and white highlights on blue paper, 8 1/8 × 7 5/16 in. (20.7 × 18.5 cm). Germanisches National Museum, Nuremberg.

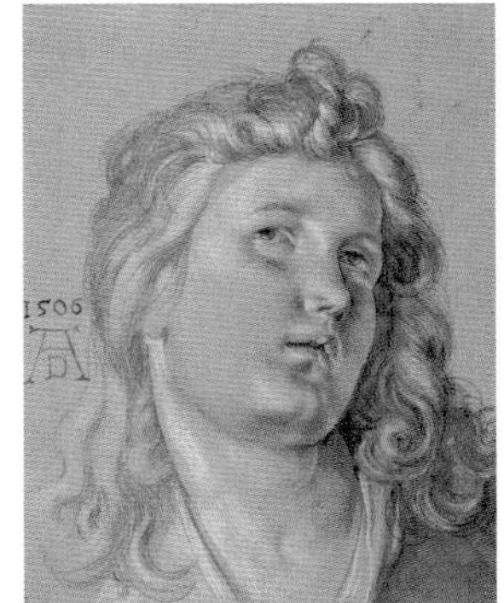

Figure 74. Albrecht Dürer, *Head of the Lute-Playing Angel from "Feast of the Rose Garlands,"* 1506. Brush drawing in gray and black with gray wash and white highlights on blue paper, 10 5/8 × 8 3/16 in. (27 × 20.8 cm). Albertina Museum, Vienna.

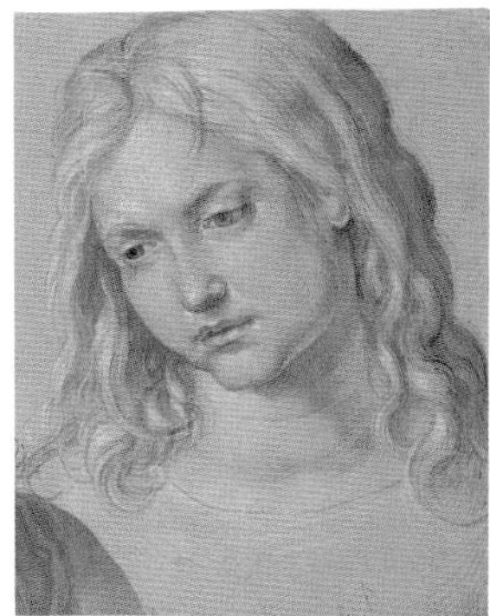

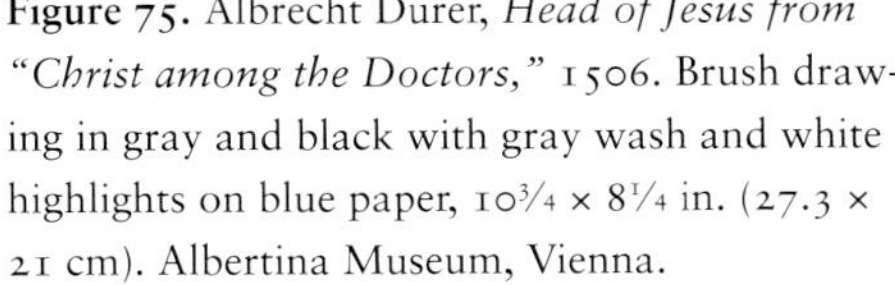

Figure 75. Albrecht Dürer, *Head of Jesus from "Christ among the Doctors,"* 1506. Brush drawing in gray and black with gray wash and white highlights on blue paper, 10 3/4 × 8 1/4 in. (27.3 × 21 cm). Albertina Museum, Vienna.

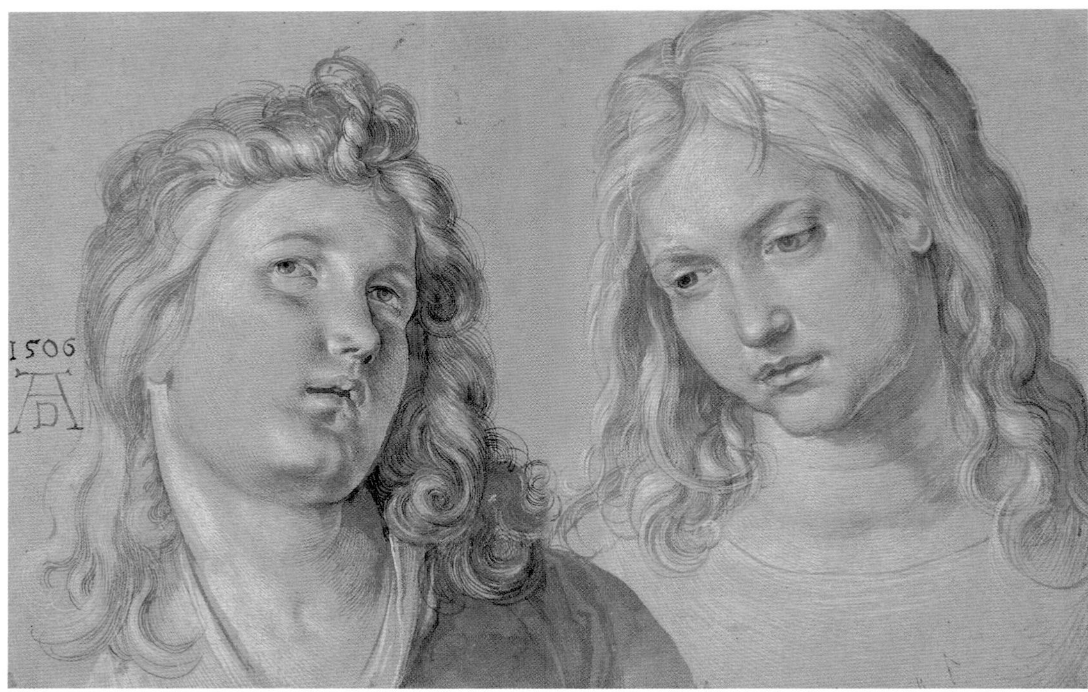

Figure 76. Albrecht Dürer, *Head of the Lute-Playing Angel from "Feast of the Rose Garlands"* and *Head of Jesus from "Christ among the Doctors"* (the two drawings reconstructed as a single sheet), 1506. Brush drawing in gray and black with gray wash and white highlights on blue paper. Albertina Museum, Vienna.

differentiated from one another."[18] Although Dürer's drawings on blue paper exist primarily in a blue-hazed monochromatic domain, the ground of the paper provided the middle tone that allowed Dürer to study how forms bend away and toward the eye. In his drawing of hands (see fig. 73), Dürer contrasted watery gray wash to bright highlights of opaque white paint. The viewer does not see white versus gray, or as Dürer would say, "how the colours are differentiated from one another." Instead, with the blue ground as a middle tone, Dürer modulated the transitions from light to dark so that the book and the hand both emerge and recede from the viewer, an impression realized by the careful juxtaposition of varied tonalities.

The blue-paper drawings offered visual evidence to his claims that he mastered color in the *Feast of the Rose Garlands*. In one of the few letters that survives from the painter's correspondence with his friend Pirckheimer, he famously reports that with the *Feast of the Rose Garlands,* he "shut the mouths of the painters who said I was good at engraving but had no idea how to use colours in painting."[19] Although the painting stayed in the Veneto, the carta azzurra drawings returned to Nuremberg as exempla of Dürer's triumph over coloristic effect. The blue paper allowed Dürer to capture in a mobile medium the technical mastery of variegating chromatic fields that he achieved in his Venetian Altarpiece.

Yet Dürer not only contextualized his Venetian work within a discourse on color, but also in terms of speed. In the self-portrait that Dürer included in *Feast of the Rose Garlands* (fig. 77), he displays an inscription that boasts about the amount of time it took for him to execute the work: "Albrecht Dürer the German accomplished this in the space of five months in 1506" (*Exegit quinque mestri spatio Albertus Dürer Germanus M D VI*).[20] Privately, in his correspondence with Pirckheimer, Dürer lamented the time it took to execute the painting, noting, "I have won a lot of praise for it but not much profit. In the time it took me I could have earned a good 200 ducats.[21]" Yet in his public display he demonstrates artistic virtuosity through the expeditious execution of the work.

In this focus on speed, Dürer solicited a comparison with Italian facility (*facilità*) and grace (*grazia*).[22] The attention to the relatively short duration of time in which the work was completed suggests that Dürer wanted to position it within a rhetorical tradition of "effortless art," a trope founded in antique rhetoric: *ars artem celare est* ("it is art to conceal art"). This expression plays out in the historiography of Italian painting between Raphael's facilità and Michelangelo's *difficoltà*.[23] Although he is famous for his difficoltà, Michelangelo repeats this language in Francisco de Holanda's *Diálogos,* as he extols the ability "to do the work that it may appear, although laboured over a great deal, as if it had been done almost hurriedly and almost without any work—and quite effortlessly, even though this is not the case."[24] Pliny the Elder also recounts the story of Pausias of Sicyon, who "being determined to give a memorable proof of his celerity of execution, he completed a picture in the space of a single day, which was then called the 'Heresios,' representing the portrait of a child."[25] Pointedly, the only other work in his oeuvre that Dürer boasted about completing quickly was his Venetian *Christ among the Doctors.* Where the *Feast of the Rose Garlands* took five months, *Christ among the Doctors* was finished in five days. On a *cartellino* falling out from one of the books held by a doctor in the foreground (fig. 78), Dürer informs the viewer: "The work of five days" (*AD opus q[u]inque dierum*).[26]

With his inscriptions regarding the relatively brief duration of their making, Dürer introduces his Venetian works into a paradigm of facility and speed as

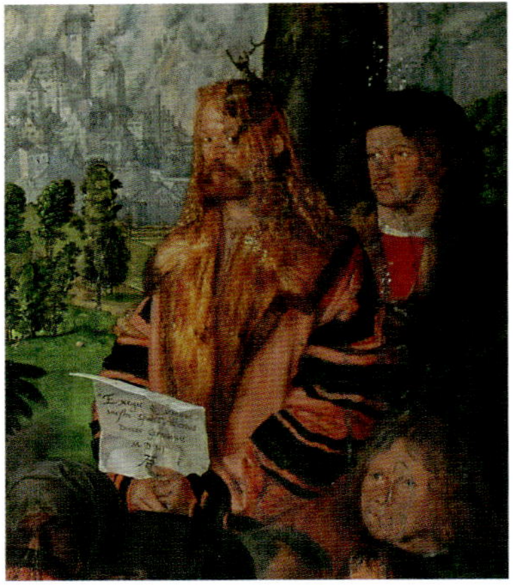

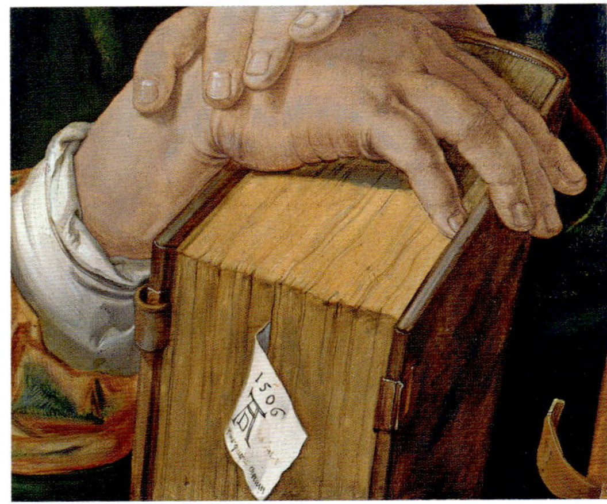

Figure 77. Albrecht Dürer, *"Feast of the Rose Garlands"* (detail).
Figure 78. Albrecht Dürer, *"Christ among the Doctors"* (detail).

marks of a master. Virtuosity and the handling of color were intrinsically connected with speed. Yet the legends within the paintings are not the only artifacts that frame these works within an antique rhetoric of "ars artem celare est." The "equivalents," to borrow Panofsky's phrase, that Dürer created in tandem with the paintings also form a category of performative drawing that enfolds the work into a discourse on speed and facility.

This was a swiftness inherent to carta azzurra that did not demand the preparation of the ground. Artists could rapidly draw on the pigmented surface. The accidents of the papermaking process become intrinsic to the drawing; the blue paper created a landscape of minute hair, fibers, and accidents of the pictorial surface as a field against which the artist realized his work. Inherent in carta azzurra was a potential effortlessness bound up in the way it was made: with a pigmented ground embedded in the sheet, the artist could work quickly and seemingly without effort. In this way, the blue ground of carta azzurra was different from a prepared ground, in which the paper is more like the ground of a panel.

To make drawings with a prepared ground, Cennino Cennini advised taking animal bones and placing them into the fire until whiter than ash, and to then "grind them as much as you can bear to grind them because they can never be done too much, since the more you grind them the more perfect the preparation comes out."[27] In this bone substance, artists would mix earth pigments to create a tinted ground and combat the chalky white surface. This powder was then mixed with gum water, which was then applied on paper or parchment with a broad bristle brush, carefully layering thin layers of the paste. Once the layers of the bone substance dried, the artists scraped with a knife any roughness from the surface before burnishing and polishing it with a boar's tooth. In the ground, the foundation is formed by the artist's own preparation from the strokes of the brush or the smooth polish attained by burnishing (figs. 79, 80). The geographic origins of the paper or the qualities of the surface were irrelevant, as the artist controlled the aesthetic effect of the ground.

In Dürer's Venetian drawings, he did not cover the paper with his own preparation, but allowed the specificity of the blue-toned support to flicker in tone, hair, and fiber against his own painterly strokes and white, heightened wash. The only works Dürer established within a vocabulary of technical mastery judged by speed were the paintings he made in Venice, for which he completed the studies on carta azzurra. While the paintings would have stayed in Venice, Dürer could bring back to his workshop the carta azzurra drawings, demonstrations of his mastery over specific concepts of artistic production: color, facility, and speed.

With blue paper, Dürer relied on local materials to manipulate his own perception and creative abilities, an experiment that would spread from his draftsmanship to his printmaking. Panofsky was one of the first to remark that after Dürer returned from Venice, the tonality of his printmaking shifted. Panofsky describes Dürer's earliest woodcuts as "an aesthetic transformation of material paper into incorporeal light."[28] Yet in his post-Venice woodcuts, Dürer transformed the paper into a gray surface with closely parallel strokes, making a middle tone to vacillate between the white of the page and deepest darkness of blank ink.[29] To draw out Panofsky's observation, Dürer's woodcuts pre- and post-Venice rely on different relationships between ink and paper, line and form. In the *Presentation of Christ in the Temple* (before Venice), the bright white of the altar is represented by the naked paper (fig. 81). This brief expanse bridges the divide between the Virgin and the priest, who holds Christ over the table, his body casting a shadow on the white cloth/paper. It is the white surface around which Dürer constructed this work, exploiting its potential to bring light into the composition. The white paper is not only the ground for the image but also becomes a figure within the composition.

In marked contrast, in the *Assumption and Coronation of the Virgin* from 1510, Dürer established the gray middle tone discussed by Panofsky (fig. 82). Although he allows moments of paper-white highlight, gray tones dominate. Dürer used the gray tones to establish a middle ground against which forms

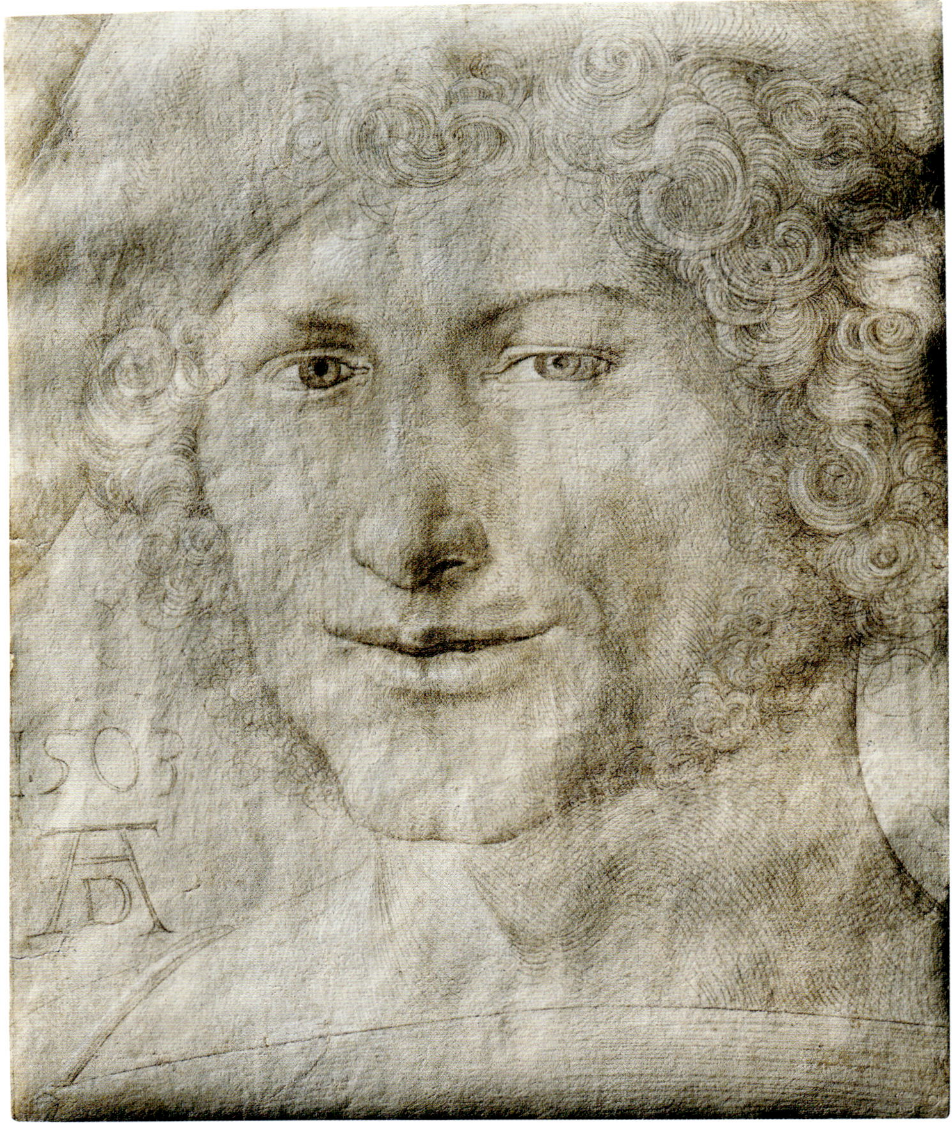

Figure 79. Albrecht Dürer, *Head of a Young Man* (image taken in raking light), 1503. Metalpoint with pen and brown ink, heightened with white on gray prepared paper, 8¾ × 7⅛ in. (22.3 × 18.1 cm). Clark Art Institute, Williamstown, Mass.

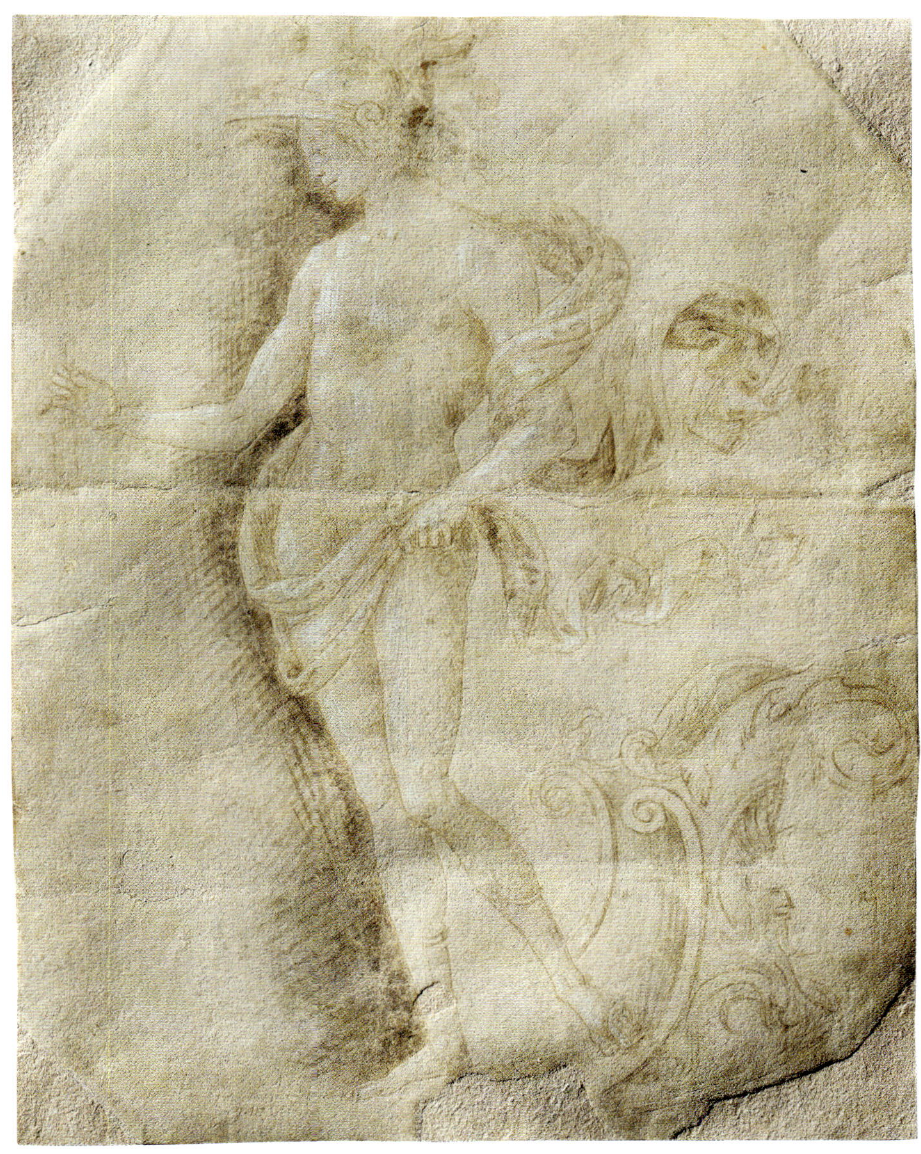

Figure 80. After Perugino, *Mercury* (image taken in raking light), ca. 1466–1600. Metalpoint, brush, and gray ink, heightened with white on light grayish-brown prepared paper, $9\frac{1}{4} \times 7\frac{3}{16}$ in. (23.5 × 18.3 cm). Clark Art Institute, Williamstown, Mass.

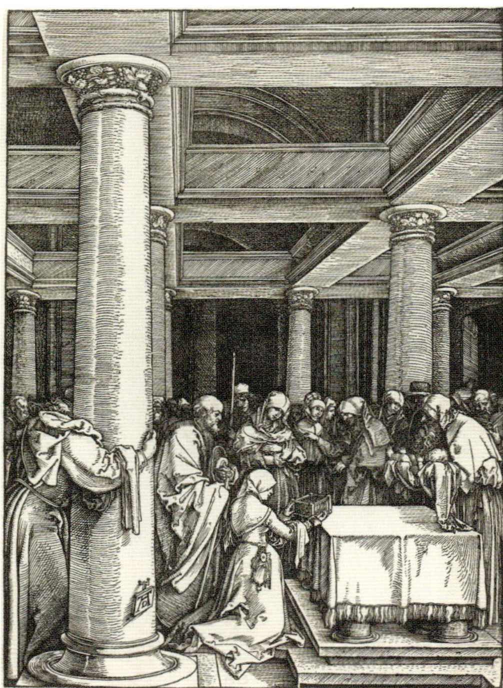

Figure 81. Albrecht Dürer, *Presentation of Christ in the Temple*, 1504/5. Woodcut on paper, 11 11/16 × 8 5/16 in. (29. 7 × 21.1 cm). Rosenwald Collection, National Gallery of Art, Washington, D.C.

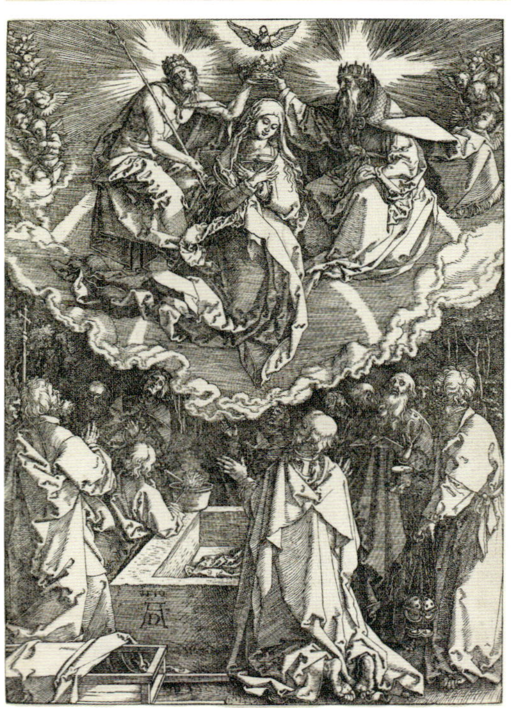

Figure 82. Albrecht Dürer, *Assumption and Coronation of the Virgin*, 1510. Woodcut on paper, 11 5/8 × 8 1/4 in. (29.6 × 20.9 cm). Rosenwald Collection, National Gallery of Art, Washington, D.C.

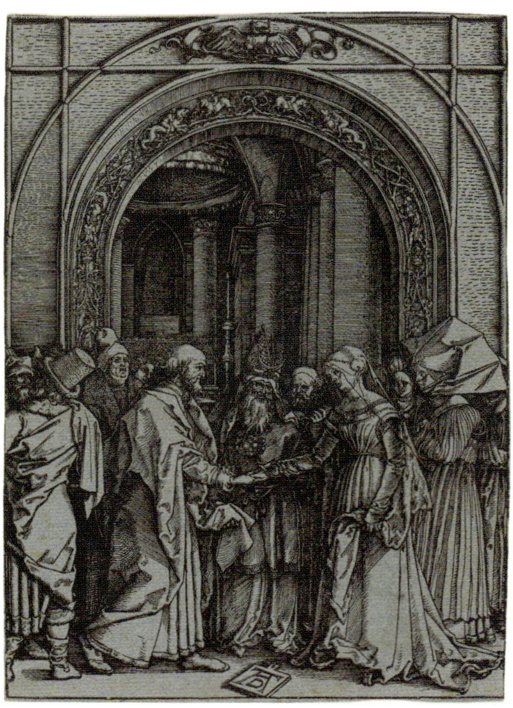

Figure 83. Albrecht Dürer, *The Betrothal of the Virgin*, 1504/5 (printed 1560s/70s). Woodcut on blue paper, 11 3/4 × 8 5/16 in. (29.9 × 21.1 cm). Ailsa Mellon Bruce Fund, National Gallery of Art, Washington, D.C.

recede and project, similar to the middle tone offered by the blue paper. Dürer did not exploit the ground of the paper as he did with the blue Venetian paper. Instead he created the ground through his own labor as a printmaker. To achieve this effect, Dürer made lines tighter and parallel to one another, also increasing the use of hatching. Whereas he established his previous composition in the white of the paper, in this later work he carved the ground of the composition. Paper is still the support for the image, but Dürer manipulated the woodblock so that it would be defined by his own work, rather than paper's contingent qualities. With this ground that established the middle tone for the image, Dürer gained greater control over the quality of later impressions, perhaps even ones struck after his lifetime.

In the pre-Venetian woodcuts of the *Life of the Virgin,* a high-quality, bright-white paper was integral to their successful execution; against this paper, the black inks form striking contrasts, dramatizing the narrative into a juxtaposition of light and dark. If Dürer created the ground himself for his woodcuts with the middle tonality, the whiteness or quality of the paper was less vital to its success as an image. The importance of paper's tonality in the legibility and execution of Dürer's works is illustrated in Dürer's pre-Venetian *Life of the Virgin* woodcut series, which was reprinted on blue paper in the mid-sixteenth century (fig. 83). The publication of the series on blue paper likely helped to increase its status as a collector's item, despite the posthumous striking of the blocks. And it is also possible that the blue paper resonated among collectors who knew of Dürer's carta azzurra drawings. Yet Dürer did not originally construct this woodcut series around a middle tonality of blue ground, and it therefore dims the force of the original images. Although it stands out as a curiosity (and perhaps distracts from the loss of legibility in the matrices themselves), this later set of imprints demonstrates that the color of paper was central to a print's impact. In creating his own middle tone in his post-Venetian woodblocks, Dürer dominated the paper by carving his own ground. Dürer's attention to building a middle tone in printmaking also demands that the viewer appreciate the effort of the artist, exhibiting a move away from the concept of facilità that he made evident in Venice. Instead, the imprinted middle tone indicates a conscious framing of his work within other tenets of artistic production, namely labor.

Drawings and Labor

Scholars often consider Dürer's Venetian drawings on carta azzurra and his Nuremberg drawings on prepared grounds in the same category, as they are all executed on a pigmented surface. It is postulated that Dürer took advantage of the local paper stock in Venice. When he returned to Nuremberg and was

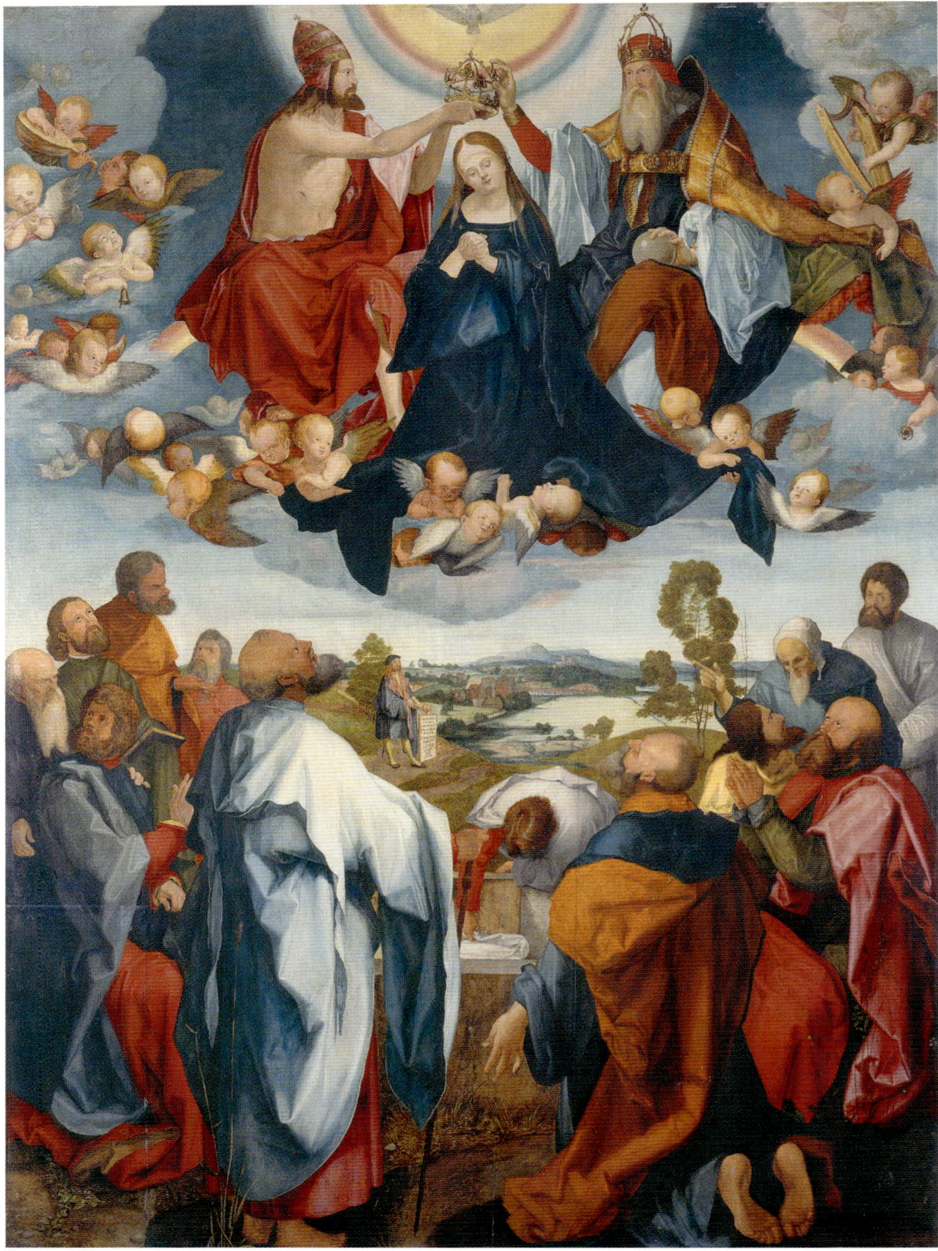

Figure 84. Jobst Harrich (after Albrecht Dürer), Heller Altarpiece, 1614. Oil paint on panel, 74 7/16 × 54 3/4 in. (189 × 139 cm). Historisches Museum, Frankfurt.

no longer able to obtain carta azzurra, he executed drawings on a sometimes blue prepared, sometimes green prepared (with red intonations) paper for the Heller Altarpiece, the central panel of which survives in a seventeenth-century copy (fig. 84).[30] Yet considering the well-established trade networks between Nuremberg and Venice, Dürer could have obtained carta azzurra in Nuremberg, and he also could have carried a stock back over the Alps, had he considered it integral to his practice. Although these two sets of drawings are on pigmented surfaces that create a tonal middle ground, the actual ground of the drawings remains fundamentally different. The accidents of the papermaking process are inherent in the blue paper, defined by the textures of fiber interwoven into both the paper slurry and impressed on its surface. When Dürer drafted on this paper, he allowed the contingencies of the papermaking process to become integral to his own contours. In contrast, the drawings on prepared ground demanded the time-consuming preparation of the ground. As Heinrich Wölfflin noted, the drawings for the Heller Altarpiece represent "Dürer more as a diligent and painstaking artisan (*Kläubler*) than as the transmitter of a direct impression of form."[31] And to this day, scholars cite the drawings as evidence of the "painstaking precision" that Dürer took to execute the altarpiece.[32]

In one drawing, Dürer distilled the Ascension pictured in the Heller Altarpiece into hands in prayer balanced in a blue void. Dürer captured the initial moment when two hands meet: fingertips lightly touch as the joints articulate away from one another (fig. 85).[33] The fingers are outlined with a thin, gray line with white highlights and hatching against the blue ground. In contrast to the soft yet firm use of line to circumscribe and shadow, Dürer deployed transparent washes of gray so that the hands dissolve into the ground of blue paper. In the uppermost hand, the gray wash drips from the cuff, softening the outline. The hands belong to an apostle gazing reverently upward at the Assumption of the Virgin, in the altarpiece's central panel. Yet in the drawing, the gesture of prayer is suspended in a blue ground. Dürer translated the dominating force of blue in the altarpiece—from the Alpine landscape to the Virgin's robe against the cerulean sky—into the opaque density of the blue-prepared ground for the drawing.

The practice of creating monumental drawings that recorded central passages of paintings became pivotal to Dürer's work following his stay in Venice.[34] Like his Venetian drawings, the *Praying Hands* was not originally conceived as a single composition. The paper was cut in two by later collectors, and the hands have taken on a force of their own in popular culture, severed from their conjunction with the bearded apostle (fig. 86).[35] Yet on the original sheet, Dürer placed the hands in relationship to a reverent, upturned face, reducing the apostle

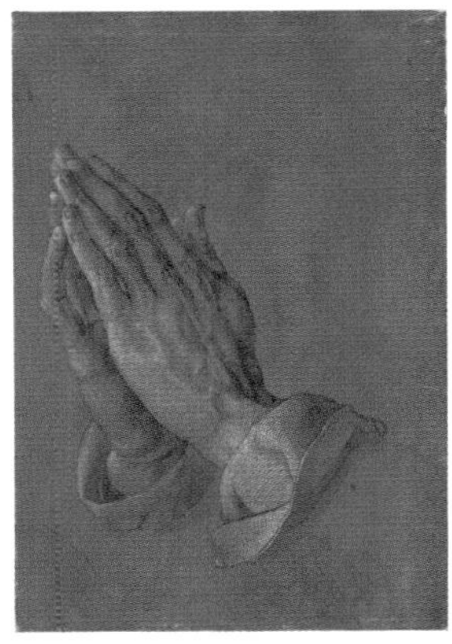

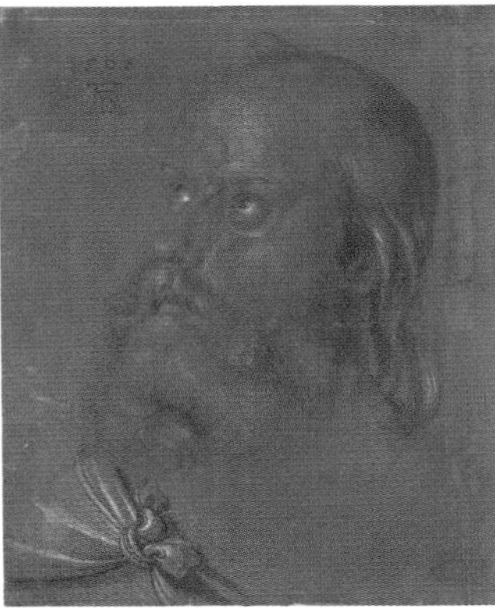

Figure 85. Albrecht Dürer, *Praying Hands*, 1508. Gray and black brush drawing with gray wash and white highlights on paper prepared with a blue-green ground, 11½ × 7¾ in. (29.1 × 19.7 cm). Albertina Museum, Vienna.

Figure 86. Albrecht Dürer, *Apostle Head Study*, 1508. Gray and black brush drawing with gray wash and white highlights on paper prepared with a blue-green ground, 12 7/16 × 9 in. (31.6 × 22.9 cm). Albertina Museum, Vienna.

to his head and hands (fig. 87). The bearded face offers a physiognomy with which to associate the hands, the delicate, veined fingers contrasting with the balding, pious figure. This type of composition is not new; a composition from the Rogier van der Weyden workshop survives in which Saint John the Baptist's profile is juxtaposed against a pair of ungainly legs (fig. 88). The drawing records Van der Weyden's *Life of Saint John the Baptist,* breaking down the painting into a gestural archive.

Both the *Feast of the Rose Garlands* (made in Venice) and the drawings for the Heller Altarpiece (sent to Frankfurt) remained outside Nuremberg. Yet with these drawings, Dürer created a record that distilled the commissions to pivotal emotive gestures. The drawings allowed Dürer to focus attention on the features of the painting that he considered integral to their reception. For these drawings were not only ricordi of Dürer's paintings, but also vital to the mythology and epistolary correspondence surrounding the paintings. The importance of Dürer's drawings in framing narratives about the paintings becomes particularly clear in the Heller Altarpiece and its accompanying sketches, letters, and fragments.

When the Archduke Maximilian bought the Heller Altarpiece in the seventeenth century, central to the acquisition was the accompaniment of the letters exchanged between Dürer and the merchant Jakob Heller. Although Heller's letters no longer survive, the remaining correspondence reveals Dürer's struggle to define the relationship between labor and economic compensation.[36] When the minor seventeenth-century landscape painter Friedrich von Falkenburg was commissioned to make a copy, he reported that it took him six months to finish.

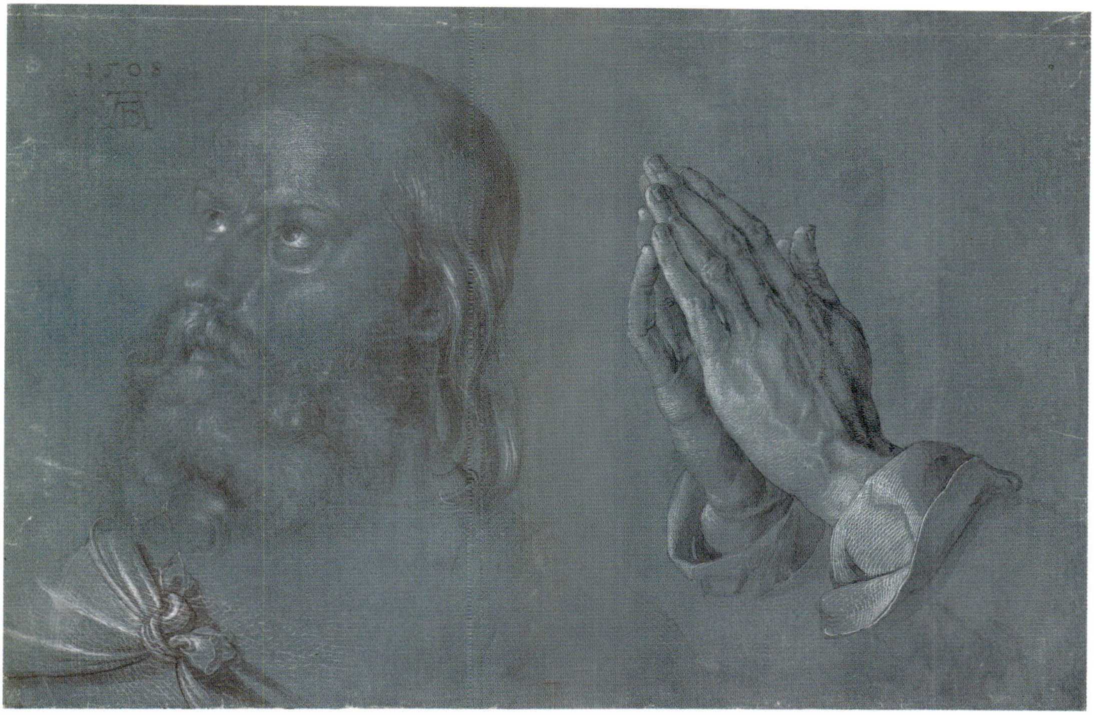

Figure 87. Albrecht Dürer, *Praying Hands* and *Apostle Head Study* (reconstructed as a single sheet), 1508. Gray and black brush drawing with gray wash and white highlights on paper prepared with a blue-green ground. Albertina Museum, Vienna.

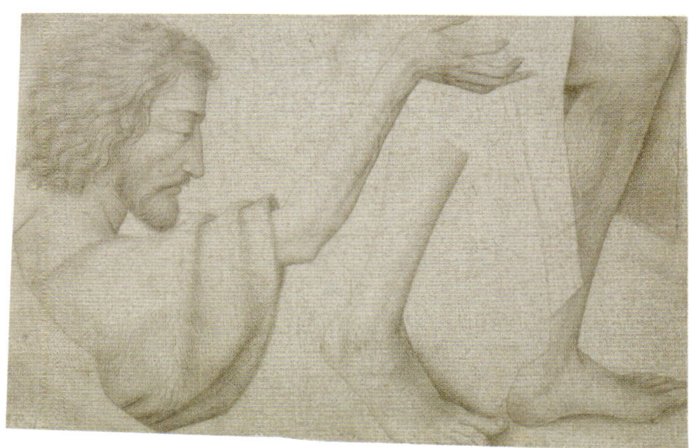

Figure 88. After Rogier van der Weyden, *Studies of Saint John the Baptist*, after 1455. Metalpoint on gray-prepared paper, $4\frac{5}{8} \times 7\frac{1}{8}$ in. (11.8 × 18.1 cm). Robert Lehman Collection, Metropolitan Museum of Art, New York.

His interlocutor expressed surprise at this protracted time, and Von Falkenburg responded that according to Dürer's letters, it had taken him thirteen months to paint the original.[37] Therefore the epistolary exchange between Dürer and his patron was essential not only to the reception of the painting, but also to the work of later copyists. And as the letters from Dürer to Heller make clear, Dürer inserted the Heller Altarpiece not within a discourse of effortless art, but rather of the labor of the artisan.

As Dürer wrote on March 21, 1509: "I have applied much effort to this one and only piece of work. There is not very much that I feel able to tell you about it, except that I am confident you will see for yourself how much trouble I have devoted to it."[38] And then again on July 10, 1509, Dürer tells Heller: "I can write to you in all truth that I have been working hard and without break on the panel and have undertaken no other work than this. It may be that I might have long since finished it, had I been willing to hurry it. But my thoroughness was meant to please you and to win praise for myself."[39] When Heller and Dürer had a dispute over time and labor, Dürer wrote to the merchant: "The painstaking care I specified, I have indeed devoted to your panel." Whereas the Venetian paintings were contextualized within a rhetoric of speed, here Dürer stressed the extensive amount of time required to complete the work: "At the same time I can't see myself painting the centre panel from start to finish in less than thirteen months."[40] As Jeffrey Ashcroft points out, Dürer engages in his letters with an established mode of exchange in Germany, in which a panel painting was on par with other artisanal crafts for which the patron paid the artist based on time.[41]

In his efforts to explain his protracted work on the panel to Heller, Dürer emphasized the preparation of the wood's ground. As Dürer tells Heller in a letter dated August 24, 1508: "And the main panel I have set out with the utmost care, taking a long time over it, and it has been undercoated with two very good layers of colour, so that I am now starting to underpaint it."[42] The undercoat provided a base tone for the panel, over which Dürer would have applied the underpaint, which worked out the layering of shadow, light, and form for the composition. As he reports to Heller: "For I intend, so soon as I hear that you approve, to paint the ground some four, five or six times over, for clearness' and durability's sake, using the very best ultramarine for the purpose that I can get."[43] Dürer's attention to the painting's ground, to assure the quality and durability of the final work, continues as a theme in the correspondence, as the painter guarantees his patron that he applied both undercoats (to prime the surface) and underpaint (for the compositional layout and coloristic effect).

The repeated application of paint as a mode of securing the worth of one's work continues in Dürer's final letters to Heller, in which he informs the merchant how to conserve his panel. As Dürer prepares the panel for shipment, he tells Heller once more: "For I put all my skill (*Fleiß*) into it, as you will see, and it's painted with the finest colours I was able to get. It is underpainted, overpainted, and fine-finished with good quality ultramarine, five or six times over. And even after the final coat I went over it again twice more, so that it would last a long time."[44]

To accentuate the time taken to complete the panel, Dürer emphasizes the painting's ground. The weight Dürer gives in his letters to the layers of paint and the preparation of the panel stresses the time required to complete the work. In turn, Dürer produced a series of drawings on prepared paper in tandem with the altarpiece. The attention to the prepared ground of the drawings reflects the repeated attention to the painting's finish that Dürer underscored to his patron. The drawings provide another form of evidence of the labor that Dürer invested in this work. They are not drafted on the raw surface of the paper, but rather are laboriously primed with multiple layers of paint, a preparatory process that mirrors his production of the panel.

The Ground of the Work

When Dürer composed his drawings, their value had yet to be established. In northern Italy and Germany, drawings were collected as containers of economic value, yet these practices were still sporadic circa 1500. As Hans Tietze remarked, drawings were valued by collectors more as "curiosities" for the *Kunstkammer* rather than as traces of an artist's intellect or hand. In his study of Dürer's drawings, Tietze argued that the Nuremberg draftsman was the first to recognize the independence of drawing as a medium instead of simply *Arbeitsmaterial* for the workshop.[45]

In his drawings, Dürer took the material of the workshop and made it a force for historicizing the reception of the artist. This was a process that he started when he collected Schongauer's drawings and inscribed them with names, dates, and biographical addenda. In turn, Dürer made drawings that would frame and underscore central tenants of his paintings, so that other artists could archive them in the future. The drawings discussed in this chapter are intrinsic to the paintings with which they were made, and therefore not autonomous. And yet they also were not designed to be constantly recycled and reused as patterns in the workshop.

Dürer employed the particular ground of his drawings and the qualities of the paper to structure his painted works and to attend to different geographic modes of making. The raw blue of the carta azzurra versus the labored brush-strokes in the pigmented ground of his Heller Altarpiece drawings represent different ideologies of craft, authorship, and the artist. The carta azzurra drawings carried over the Alps the mode of making that Dürer explored in Venice: speed, color, and facilità. In turn, the drawings for the Heller Altarpiece engage with a Northern dialogue of labor, time, economic compensation, and the artisan. Through paper, Dürer technically engaged with different ideas of authorship.

Dürer's Venetian and Heller Altarpiece drawings stand out because their reception was not only founded on their iconographic vocabulary but also their material qualities. One of Dürer's earliest copyists, Hans Hoffmann (1530–1591), attended to the importance of the pigmented preparation of the paper when he made a drawing after Dürer's *Head of an Apostle,* copying both Dürer's study and the material techniques that Dürer employed to execute the original drawing (fig. 89). Hoffmann therefore not only imitated Dürer's iconography but also his preparation of a body-color ground. Moreover, like Dürer's annotation of Schongauer's drawings, Hoffmann inscribed Dürer's monogram and the date of the original: 1508.

Similarly, when Aegidius Sadeler (1570–1629) translated the carta azzurra drawings into engravings, he maintained the sculptural quality of the drawings (figs. 90, 91). Only instead of using blue paper to create a middle tone, Sadeler emulated the blue paper with densely parallel lines. Where Dürer applied opaque white paint for highlights, Sadeler utilized the white paper to illuminate the figure. He translated the fine strokes of Dürer's wash into burin lines, creating a similar effect that Dürer created with brush and pigment. Yet Sadeler also departed from the original design, adding a backlit shadow to both figures and Dürer's monogram in the upper-right corner. The shadow cast on the wall by both figures indicates that the original sheet of paper was already divided in half when Sadeler made his engravings. With the strong presence of the shadowed wall, it is impossible to conceive of the two faces sharing the same ground, as they are now incised into their own paper niches. The inclusion of the shadow detaches them from the floating blue surface on which they were carefully placed and makes them figures in relief. Sadeler also included an inscription in Latin: "*ALBERTUS DURER ALMANUS FECIT ANNO MDVI / EGIDUS SADELER SCALPSIT ANNO / MDXCVIII*" (Albrecht Dürer the German made this in the year 1506 / Aegidius Sadeler engraved it in the year 1598). Like Dürer's annotation of the Schongauer drawing with both the years 1470 and 1517, Sadeler demarcates two points in time: the year the original image was made, and the year

Figure 89. Hans Hoffmann (after Albrecht Dürer), *Head of an Apostle*, 1576–84. Brush drawing with pen and black ink, heightened with white on blue-prepared paper, 11 9/16 × 8 7/16 in. (29.3 × 21.4 cm). British Museum, London.

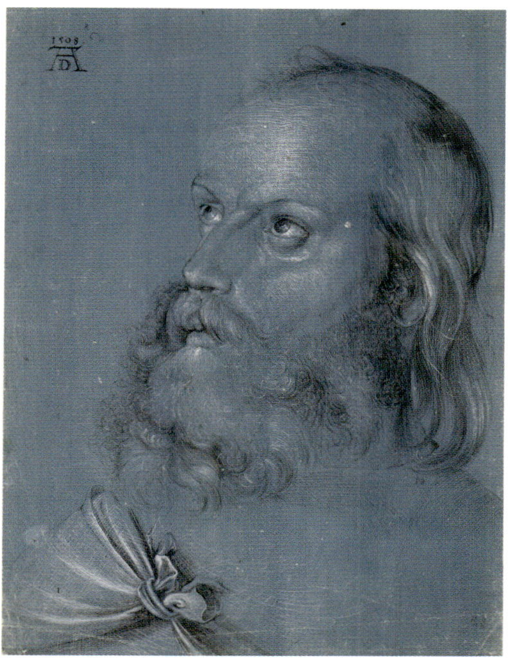

Figure 90. Aegidius Sadeler (after Albrecht Dürer), *Head of Lute-Playing Angel*, 1598. Engraving on paper, 14 1/16 × 8 15/16 in. (35.8 × 22.7 cm). Rijksmuseum, Amsterdam.

Figure 91. Aegidius Sadeler (after Albrecht Dürer), *Head of Jesus*, 1598. Engraving on paper, 14 1/16 × 8 5/16 in. (35.7 × 22.7 cm). Rijksmuseum, Amsterdam.

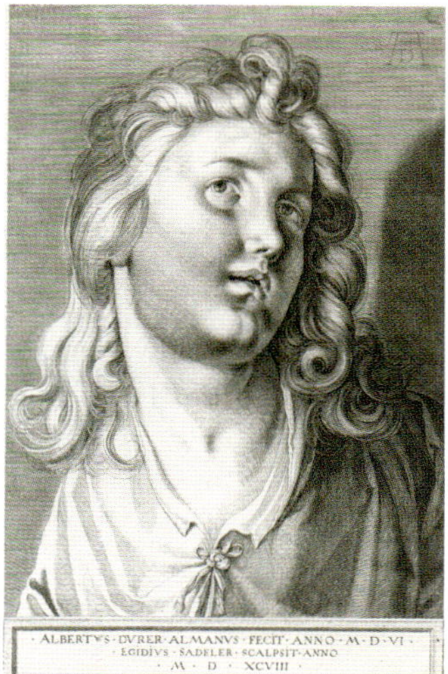

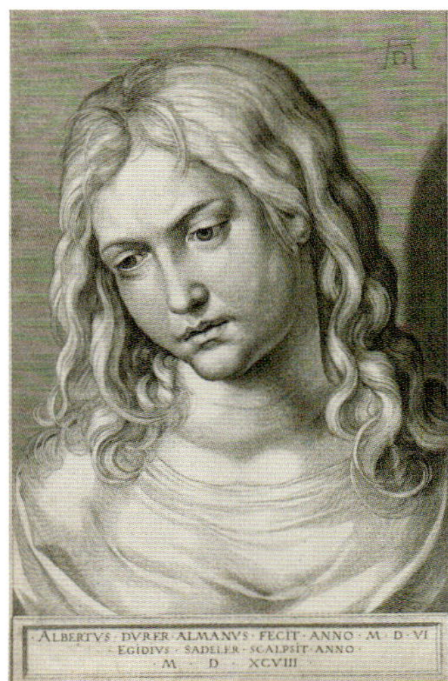

that he engraved it. Yet his inscription also departs from Dürer's, as he employs "scalpsit" in the caption to demarcate the translation of the image across media. These multiple dates coexist while also establishing a linear framework for making, reception, and influence. Paper created the mythology of a direct encounter between a dead artist and a contemporary viewer, providing a ground on which to articulate a momentary collapse of time in the awareness of its passage. This power was made present in the inclusion of two dates: 1470 and 1517, 1506 and 1598.

Early modern drawing scholars track a teleological history of drawing toward the autonomous work that exists free from any functional role in the workshop; yet the attention to the workshop model versus the autonomous drawing obscures works on paper in ambivalent states. With his term "equivalents," Panofsky disentangled this category of the Dürer workshop from the subordinated role of studies for paintings. Instead, they are *equivalent* to painting and carry in their paper grounds a history that is both integral to and distinct from the paintings with which they were made. They created a new category of work that was copied, annotated, and translated across a paper world that framed, contextualized, and mythologized the work of the painter. For it was on paper that the encounter between the past and the present was named, dated, and archived.

CHAPTER V

Paper

Modernizing the Ancients' Wax Tablet

Early modern rag paper contains a pictorial language that historians rarely address unless they want to place a work within a chronology. This idiom is called the watermark. Thanks to the exhaustive cataloguing by scholars such as Charles-Moïse Briquet in *Les filigranes: Historique des marques du papier dès leur apparition vers* 1282 *jusqu'en* 1600 (1907; *Watermarks: Historical Marks on Paper from Their Appearance around* 1282 *until* 1600), watermarks may be referenced and identified among the compendia of anchors, foolscaps, hands, horses, and unicorns. But once the date of the work is established through the watermark, historians ignore the image in favor of the text or figure on the paper's surface.

Yet if light casts its rays through the surface, the watermark becomes intrinsic to paper's body. There are moments in Renaissance drawings when there is a resonance between the watermark and the image, as in Michelangelo's red chalk drawing *Archers Shooting at a Herm* (fig. 92), where the crossbow watermark parallels with the drawing's subject (fig. 93).[1] Yet these moments of concord are rare. For the most part, overt attention to the watermark unravels the unity between the image and its support. In recognition of the watermark's potential disruption of the image, Albrecht Dürer's earliest impressions of his *Meisterstiche*—the three masterworks of engraving known as *St. Jerome in His Study, Melencolia I,* and *Knight, Death, and the Devil*—are defined by the absence of

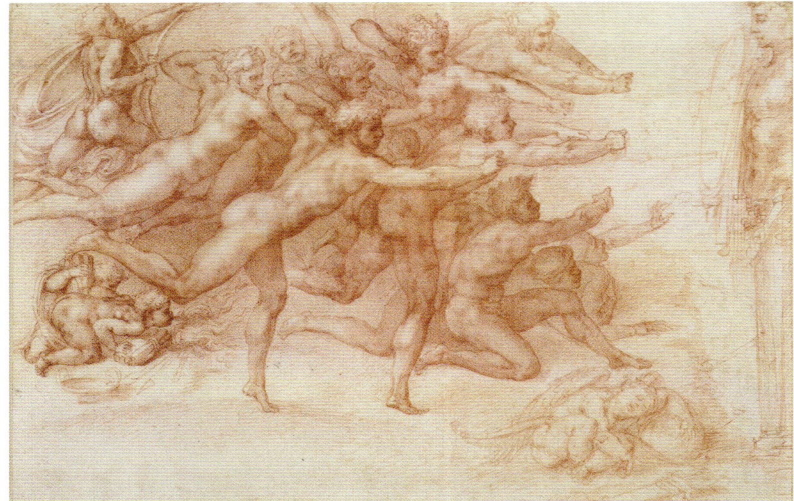

Figure 92. Michelangelo, *Archers Shooting at a Herm*, ca. 1530. Red chalk on paper, 8⅝ × 12¾ in. (21.9 × 32.3 cm). Windsor Castle, London.

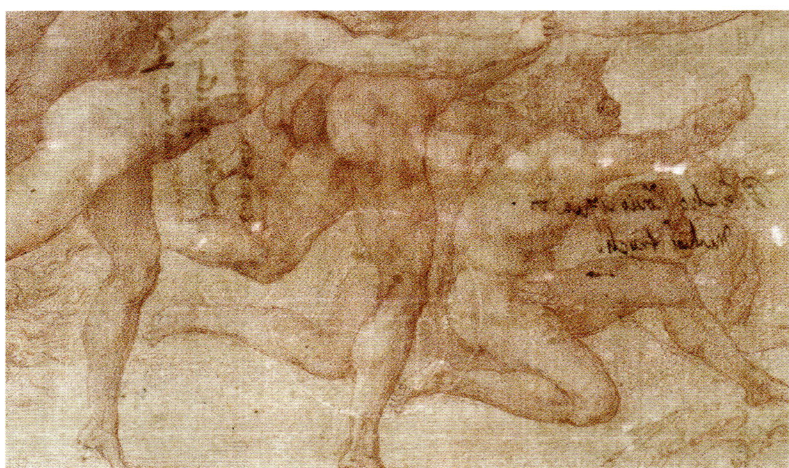

Figure 93. Crossbow Watermark in Michelangelo, *Archers Shooting at a Herm*, ca. 1530. Windsor Castle, London.

watermarks. Dürer either had paper specially made without watermarks, or he positioned the plates so as to omit them.[2] For Dürer, the watermark was a visual disruption. Yet outside of Europe, the watermark could herald a crisis in the image itself. Fifteenth-century Muslim jurists debated about writing the Qu'ran on paper with Christian watermarks of the paschal lamb, or the Greek or Latin cross, as they were concerned that these signs would stimulate idolatry.[3] This fifteenth-century legal discourse around watermarks is rare, however, and for the most part the evidence of the reception of watermarks remains sparse. Yet this pictographic language continued to present itself in documents, treatises, codices, and drawings across the early modern world.

The presence of watermarks becomes particularly palpable during the encounters in New Spain between the Spanish and Nahuas, as the conquistadores imported European paper to exact their colonial bureaucracy on the indigenous populations. European paper distinguished itself from indigenous paper with one central feature: the watermark. This sign represented a relationship to land and property embedded in the legal composition of European paper. And it was this overriding structure of land and property that provided the surface on which the colonial administrators enacted their bureaucracy.

Two Pictographic Languages

The introduction of European paper into Mesoamerica simultaneously occurred with the transition from one system of writing to another, from pictographic to alphabetic writing. The watermark, however, complicates this encounter, for it was a pictographic form of European writing that existed both outside and within alphabetic construction. While watermarks frequently incorporated letters, they also consisted of pictorial signs: hands, sheep, and crosses, a visual language that reveals an alternative literacy embedded in the sheets of European paper.[4]

In an examination of one particular work, the Codex Telleriano-Remensis, the cohabitation of the watermark and the images discloses the coexistence of two different pictographic languages. The Codex Telleriano-Remensis contains calendars and the history of the Aztecs through the conquest. It was made for a European patron by indigenous painter scribes known as *tlacuilos,* and takes its name from the archbishop in France who owned the codex in the seventeenth century. As the reader opens the manuscript, the watermarks reveal themselves within the pages featuring the extensive image cycles made by the tlacuilos. In folio 19r, the artist depicted a figure adorned in plumes of leaves and knives, known as the broken tree, which symbolizes Tamoanchan, a lost mythical site of origins (fig. 94).[5] The scrawled notes of commentators identifying the painted figure frame the central image. Nevertheless there is another picture embedded in the page for which there is no gloss: a gloved hand with the initials "BF" across the palm, crowned by a five-petaled flower (fig. 95).[6] This is the watermark. Although scholars remark on it as the sign of European paper, the watermark and the image are always reproduced separately.

There are no known records of the reception of European watermarks in Mesoamerica. And it is difficult to account for whether or not this symbolic language carried any significance for the scribe-painters as they incorporated

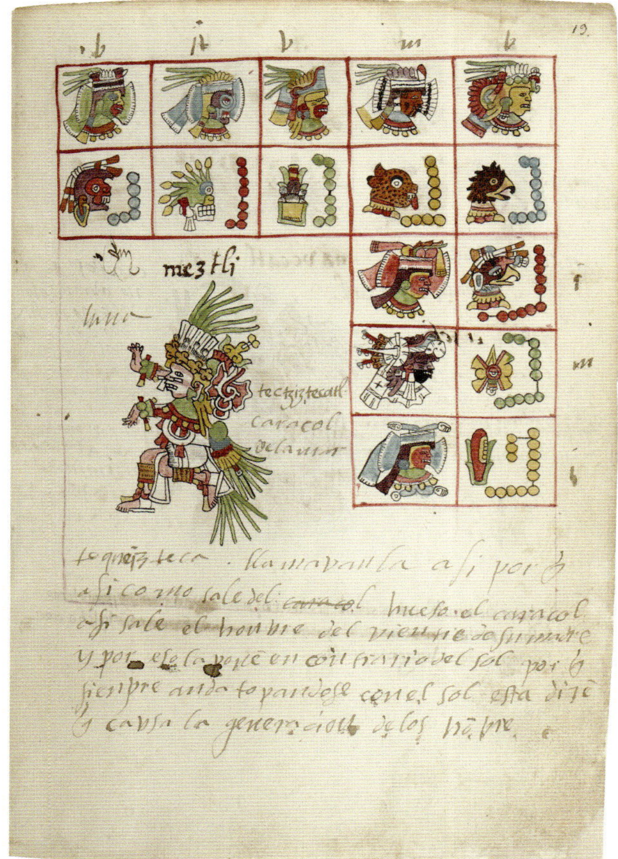

Figure 94. Fol. 19r in the Codex Telleriano-Remensis (Mexicain 385), after 1562. Opaque watercolor and ink on European paper. Bibliothèque nationale de France, Paris.

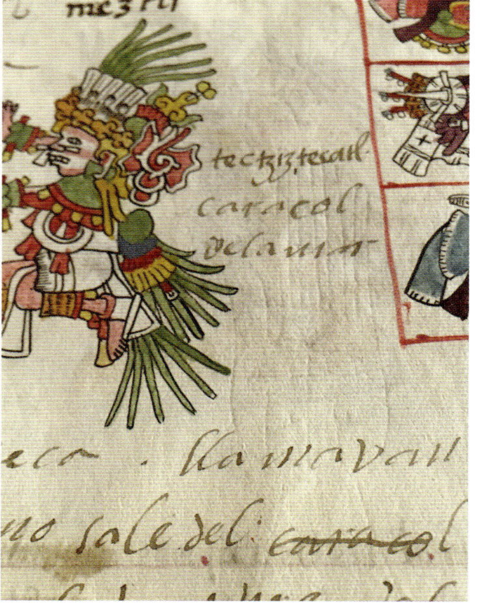

Figure 95. Gloved-Hand Watermark on fol. 19r in the Codex Telleriano-Remensis.

the foreign product into their workshop practice. Although given that the geographic specificity of paper across the empire was also denoted by a pictorial language, as seen in the *Matrícula de Tributos* (discussed in Chapter 1), the language of the watermark as a site of origins was possibly not foreign to the local scribes. Yet the watermark also transported specific ideas about authorship and property into non-European worlds.[7] As the gloved hand reminds the reader, watermarks represented larger institutional structures.[8] Paper was a ritual and archival material in Mexico since at least 600 CE, when papyrus and parchment were still the dominant materials of the Mediterranean. Yet despite the relatively short history of paper manufacture on the European continent, by the seventeenth century it was a Christianized European product and an embodiment of specific ideas about self, property, and land. The watermark was not the only visual system embedded in European paper. It also carried the grid of the wire mold impressed on the paper's surface as a network of tightly packed laid lines and chain lines—a structure absent in the indigenous paper.[9]

Water and Paper

In one of the earliest European discussions of paper, the Italian jurist Bartolo da Sassoferrato (1313–1357) examined the watermark and, specifically, whether it signified the property of an individual or a place. For Bartolo, the watermark denoted the physical environment of the mill, indicating the place rather than the artisans who crafted the paper. While he acknowledged that some "trademarks are proper to a particular craft or skill," he maintained that watermarks identify "the nature of the place in which the product is made." The watermark is the property of the mill, or "the one to whom the mill itself belongs, no matter whether it remains in his possession by right of ownership or lease, or by any other title, wholly or in part, or even in bad faith. During the entire time in which he has possession of the mill, he cannot be prohibited from using the watermark, as with other rights to immovable property."[10] For Bartolo, the watermark designates the mill's water quality. This condition of the mill's water was determined by its location, and whether it was built at the source of the running water or closer to the ocean and brackish water. As watery bodies run their course, their transport of mud, flora, and other detritus enters into the paper slurry. The mill's position in relationship to the water's course determines the debris in its running water. Other topographical features such as a spring or the surface water from rain and snow also impact the mill's water. Bartolo's recognition of the relationship between the watermark and immovable property demonstrates a sensitivity to the relationship between site and quality, an attention to local environment that continues today among artisan papermakers. For the contemporary papermaker Timothy Barrett, clean running water is fundamental: "For papermakers of any era, stray bits of foreign matter seem to enter the pulp no matter how much care is taken to keep them out. . . . Clear fresh water is, and was, of the essence."[11]

In his monumental study of watermarks, Briquet points out that many of the first watermarks identified specific papermakers. For the first thirty years, watermarks are a cross with a letter.[12] Yet in the fourteenth century, the mark of the individual virtually disappears. Instead watermarks increasingly represent the location of the mill, conveying that paper quality was not dependent on a specific papermaker, but on the mill's physical site and the purity of the water.[13] Although paper encouraged mobility with its lightness, the wire mark meant that paper would always carry what could not be transplanted: its indigenous environment.

Bartolo's discussion of the watermark in a treatise on arms and insignia reveals a telling context: the wider world of merchants' marks, prevalent since at least the thirteenth century in Europe.[14] Northern Italian trade, in particular, depended upon merchants' marks to discriminate among different producers and the quality of their commodities. The frequent occurrence of this language in local townscapes is depicted in a predella panel by Gentile da Fabriano (1370–1427), which illustrates the story of Saint Nicholas of Bari resurrecting three youths pickled by an innkeeper (fig. 96). On the walls of the inn, Gentile painted the marks of merchants, linear geometric signs that formally resonate with the emerging vocabulary of the watermark.[15] And extending the importance of these symbols beyond northern Italy, the bibliographer Allan Stevenson points out that the shops of stationers and printers frequently employed watermarks as shop signs, a visual language that structured the urban landscape of commerce.[16]

In Gentile's painting, the merchant's mark is a concrete part of local urban landscapes. It distinguished origins, producers, and the economic value of a product. When the language of the merchants' mark was incorporated into the mobile material of paper, it created a network of signs that moved beyond the city walls. Watermarks traveled among the major European papermaking centers and across oceans to demarcate economic value and origins. With the large-scale production of paper, the watermark became severed from specific landscapes of urban commerce and became a general indicator of quality and place. Yet Bartolo's treatise demonstrates how the origins of this language were founded in an attention to environment, and its importance in defining economic value.

Two Paper Epistemes

This opaque and pictographic language of trade also journeyed beyond northern Italy and into New Spain, where "knowledge and 'paperwork' went together," as the Spanish crown ran a paper bureaucracy between the Old World and the New.[17] Although there is evidence of sixteenth-century paper mills outside of Mexico City, the predominant support for New Spain's bureaucratic structure was imported Genoese paper and its indigenous counterpart.[18] The sign of

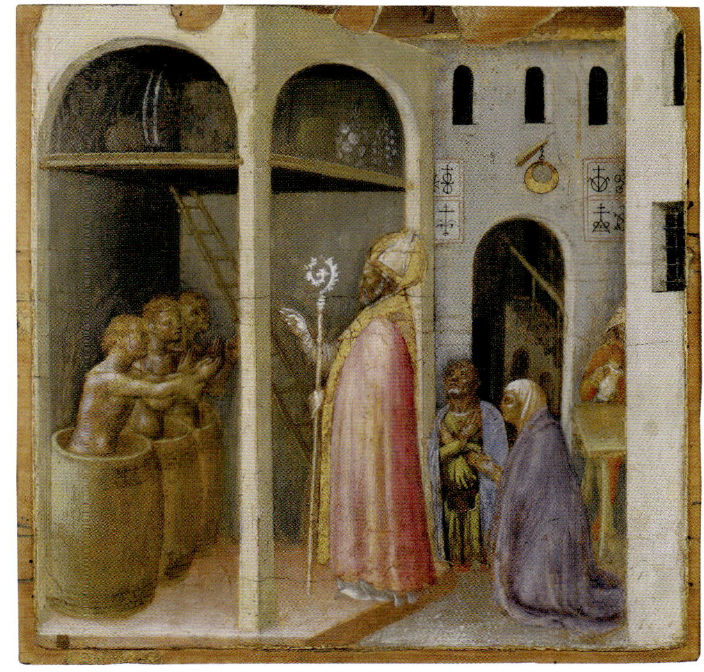

Figure 96. Gentile da Fabriano, *Quaratesi Polyptych: Saint Nicholas of Bari Raising Three Boys from the Dead*, 1425. Tempera on panel, 14 × 14 3/16 in. (35.5 × 36 cm). Pinacoteca, Vatican Museums.

Genoese paper is the gloved-hand watermark, deployed in a seemingly infinite variety of iterations throughout paperwork in New Spain. Yet both European and indigenous paper provided supports for land disputes in the New World, as litigants defended their rights to land either by invoking the antiquity of the indigenous support or entering into the European domain of law and property, expressed in its paper.[19] In her work on colonial law among the Nahuas, Susan Kellogg argues that the establishment of a legal system was formative in Spain's success; both cultures defended their rights, enacting a power struggle not only for legal recognition but also for the structure of cultural norms.[20] Paper was the material on which these battles were carried out.

One of the earliest surviving sixteenth-century legal documents from ten years after the conquest, the Huexotzinco Codex (1531), marks the presence of local paper in legal proceedings. Testifying to the excessive taxes and labor levied on the Nahuas, the painter-scribe utilized the support readily available, as European paper was expensive and likely difficult to obtain at this early date after the conquest. Typically reproductions from the Huexotzinco Codex focus on isolated images, as it is famous for having one of the earliest representations of the Virgin Mary by a tlacuilo (fig. 97), painted on a type of paper made from the agave plant known as maguey. Yet the codex is a larger historical document consisting of imported paper from Genoa and two kinds of paper: *amatl* (fig. 98) and maguey. The final document contains evidence on Genoese paper (fig. 99), and the Nahua declaration on indigenous paper. The testaments of each group obtain their authority from the support deployed to transcribe the evidence. The paper provided by the Nahuas offers a valence of authenticity that traveled beyond the colonial court of Mexico and into Spain and the court of Charles I.

Questions remain, however, regarding the two kinds of indigenous paper deployed in the testimony. Amatl was made from tree bark, while maguey was produced from agave fibers. While geographic origins and site were embedded within the discourse of European paper, it is plausible that the use of these two types of paper also articulated a specific relationship to land and place. The coexistence of the two types of paper in the legal document may be contingent, but it could also speak to a broader awareness of distinct values embedded in the varying supports. By using varying indigenous supports, the litigants could carve out a space of "knowledge and paperwork" that remained decipherable to the Nahua communities, who maintained their own orientation to land, while also making a case legible to the colonial court. Paper was a translatable commodity: Europeans understood these visual testimonies as a form of bureaucratic paperwork. And yet the paper also embodied its own technology, distinct from Europe.

Nahua litigants deployed European paper as well, as may be seen in the Codex Kingsborough, or the Codex Tepetlaoztoc. This document enumerates the abuse of local resources by the Spaniards, attesting to the exploitation of the tribute system by the conquistadores.[21] But the manuscript also bears witness to the ways in which the Spaniards transformed the landscape, specifically through building water mills with forced indigenous labor. In several folios, the tlacuilo illustrated a water mill, making evident both the structure that had been built with forced labor and the industrial impact of the Spanish on the local landscape (figs. 100–102). The rigid geometry of the rectangular bricks and the circular wheel contrast with the representation of water, conveyed through an interweaving structure of line, circle, and the articulation of wave crests into geometric patterns.[22] By illustrating their complaints against the Spaniards on European paper, the community turned the support of colonialism into a legal document directed against the Spanish. The rows of tributes and the visual accounts of the imposed water mills built with forced labor provide evidence for the exploitation of labor and industry. The presence of the European paper

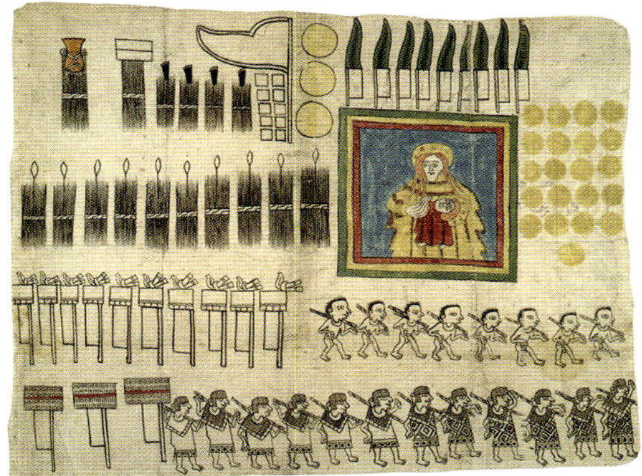

Figure 97. Image depicting tributes given to the Spanish, including a feather painting of the Virgin and Child, in the Huexotzinco Codex, 1531. Opaque watercolor on maguey. Harkness Collection, Library of Congress, Washington, D.C.

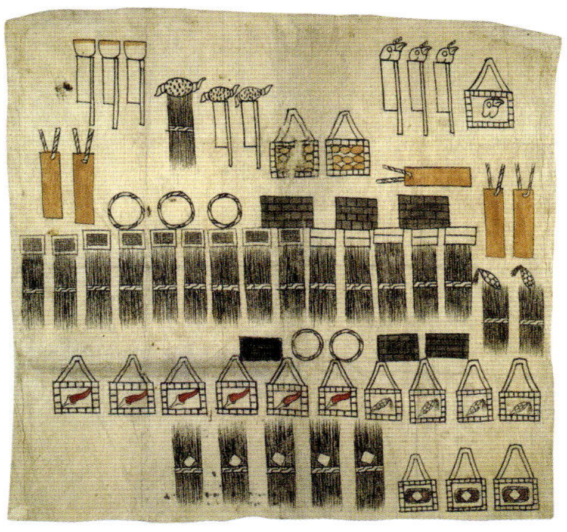

Figure 98. Image depicting tributes given to the Spanish in the Huexotzinco Codex, 1531. Opaque watercolor on amatl. Harkness Collection, Library of Congress, Washington, D.C

activated a commodity that embodied the specific industrial relationship to land, waterpower, and the misuse of agriculture against which the legal document was testifying. By using the imported product of the conqueror—a material constructed through mill power—the tlacuilos stressed the forced importation of industry onto the landscape.

Paper was a complex tactile, visual, and audible language across the documents and codices of the early modern world that evoked both distant and near ecologies. The paperwork of the colonial court of the Indies demonstrates the heterogeneity of paper in the early modern world. While the colonial court was defining a universalized idea of property, the varying supports within the colonial court communicated divergent ways to structure a community's relationship to land, self, and labor. Yet despite the many types of paper in circulation in the early modern world, an *idea* of paper as distinctly European emerged.

Paper as the New Tabula Rasa

White paper from Europe came to embody the universal idea of paper—and, metaphorically, the human mind. In John Locke's *Essay Concerning Human Understanding* (1689), he described the mind as "white paper devoid of all characters, without any *Ideas.*" Locke then rhetorically asked how this sheet

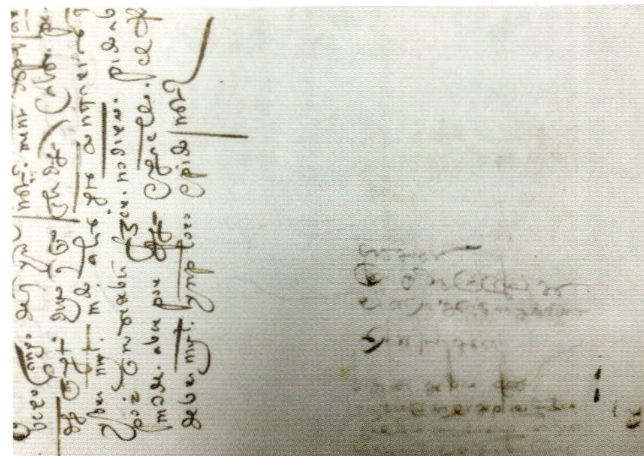

Figure 99. Genoese Gloved-Hand Watermark in the Huexotzinco Codex, 1531. Ink on European paper. Harkness Collection, Library of Congress, Washington, D.C.

becomes "that vast store, which the busy and boundless Fancy of Man has painted on it with an almost endless variety." In answer, Locke proposed that paper becomes furnished through "*Experience.*"[23] Scholars maintain that Locke modernized the ancients' wax tablet, as his likening of the mind to white paper belongs within a longer tradition of the tabula rasa. In this epistemological tradition, the individual's mind is equated to a wax blank slate at birth, which will become filled through external and internal sensation—a philosophical convention founded in Plato's *Theaetetus* and Aristotle's *De Anima* (*On the Soul*).[24]

In Plato's dialogue, Socrates states: "Please assume, then, for the sake of argument, that there is in our souls a block of wax, in one case larger, in another smaller, in one case the wax is purer, in another more impure and harder, in some cases softer."[25] This tablet forms the model of mind and memory, as impressions are imprinted on its malleable surface. Yet the dialogue also demonstrates that the variable qualities of wax explain the differing capacities of souls to create their inner worlds, as Plato discusses the changeable qualities of wax from a "pure" wax that is soft and ready for a variety of impressions to an impure and hard wax that cannot receive imprints. In Aristotle's treatise, the Platonic discussion of wax becomes a means to clarify that sense perception merely receives an imprint, and not the physical material of the object: "We must understand that sense is that which is receptive of sensible forms apart from their matter, as wax receives the imprint of the signet ring apart from the iron or gold of which it is made."[26] Despite their related yet distinct ideas about the formation of knowledge, both Aristotle and Plato utilize the traits of wax in their description of how the mind is impressed, an epistemological metaphor that would become a basis of Western philosophy.[27]

Figures 100–102. From the Codex Kingsborough/ Petition of the Indians of Tepetlaoztoc, ca. 1500. Written and painted on European paper. British Museum, London. Right: fol. 37v; below: fol 42v; below right: fol. 42r.

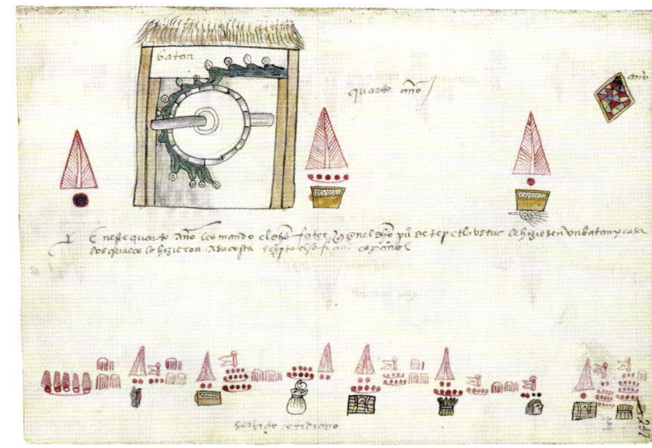

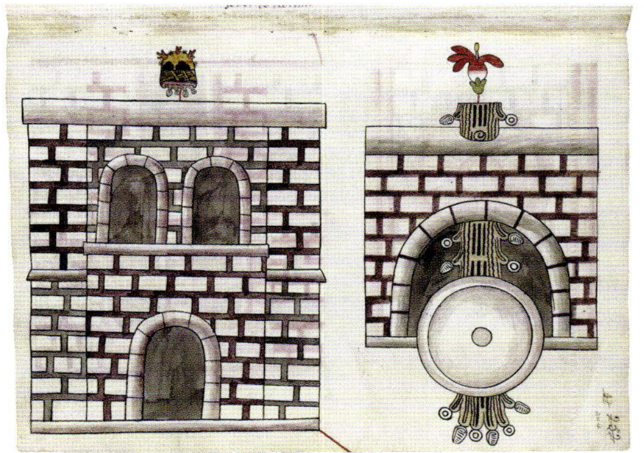

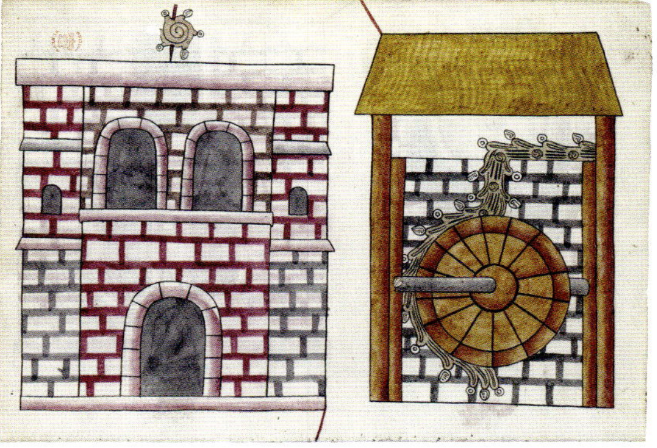

It is necessary, therefore, to consider the unique qualities of paper that made it relevant as an image of the mind in the seventeenth century. For Locke was not the only writer to liken the intellect to paper. Earlier in the seventeenth century, Francis Bacon (1561–1626) compared the mind to a "fair sheet of paper with no writing on it."[28] And Locke's contemporary, Thomas Hobbes (1588–1679), associated wits to paper when he wrote: "Whereas the Common-people's minds, unless tainted with dependence on the Potent, or scribbled over with the opinions of their Doctors, are like clean paper, fit to receive whatsoever by Publique Authority shall be imprinted on them."[29] Locke and his fellow early modern philosophers demarcated paper with potent adjectives. For Bacon it is "fair." For Locke it is "white," and for Hobbes "clean." The adjectives clarify that this paper mind is not like the popular blue paper that Dutch artists utilized for chiaroscuro drawings, or the brown paper used for wrapping merchandise and goods. Instead, the seventeenth-century philosophers stressed that the pre-sensate mind must be compared to a sheet of expensive, white writing paper.

These analogies between paper and intellect may not seem paradoxical today. Contemporary office paper is nondescript and made a bright white through optical whitener. The imitation of this paper in word-processing programs presents a glowing blank space for inscription. Even on the average sheet of recycled paper, there are no idiosyncratic textures but merely the muted shade of brown that assures ecological consciousness. Yet the paper that Locke employed was not the frictionless sheet made from wood pulp processed to oblivion. In the seventeenth century, European paper bore the traces of its linen fragments in its surface and body. Rag paper also carried the gridded indentations from the papermaker's mold and the watermark. And while it is certain that Locke's sheet was seventeenth-century European paper, vegetal supports from other places in the world would have come to his attention. Nevertheless, paper becomes abstracted into an entity with no history, no geography. It becomes a tabula rasa.

It could be argued that Locke and his fellow early modern philosophers considered the presence of watermarks and rags in their discussion of paper, and that these informed their analogy to imply traces of innate knowledge in an otherwise tabula rasa framework. Bacon concedes that the mind is not as fair and smooth as paper: "But since the minds of men are strangely possessed and beset, so that there is no true and even surface left to reflect the genuine rays of things, it is necessary to seek a remedy for this also."[30] For Bacon, the human mind remains less "true and even" than paper, as it is beset by passions and false knowledge. Moreover, seventeenth-century writers did not have at their disposal the industrial paper of today. Expensive white writing paper was fair and smooth, particularly in comparison to its cheaper counterparts. Nevertheless, it was not a tabula rasa. By attending to the qualities of paper, it may be considered how Locke's likening of the mind to paper extends beyond modernizing the ancients' wax tablet. The introduction of paper as the surface on which the world is reflected, stored, and processed introduces a shift in the articulation of sensory experience and the corresponding fields of memory, perception, and judgment.

When Locke analogized the pre-sensate mind to "white paper devoid of any *characters*," he referred to a product of refuse, transformation, and industry. Locke did not associate the tabula rasa of the mind with earth or stone. He named a made thing, a made-ness that fascinated Bacon when in his *Novum*

Organum (1620; *New Organon*) he celebrated paper for its technology. For Bacon, paper mimicked materials found in nature, while remaining distinctly artificial. He described paper as "not fragile like glass nor woven like cloth; it certainly has fibres, but not distinct threads, altogether like natural materials. And so it is found to have hardly any similarity with other artificial materials but to be absolutely unique."[31]

This construction of paper was integral to Locke's appreciation of it as a metaphorical surface for the mind; his equation of the pre-sensate mind to an already manufactured sheet of paper highlights man as a product of design. And it is because man is God's creation (a manufactured product) that he is also God's property. Locke's philosophy depends upon his concept of God as maker, which gives God dominion over man.[32] Yet this logic carries over into human law. Just as God has dominion over man, so man has dominion over the products of his own labor.[33]

Locke's writing on property in book five of his *Second Treatise of Government* (1689) explicitly engages with the property rights of colonial empires in the Americas, an interest that developed during his time as secretary for the Earl of Shaftesbury, who was invested in the Carolina colony. In book five, Locke defends his patron's rights to the Carolina land by applying the same logic of labor and dominion, arguing that America was a wasteland that the colonists transformed with their labor. Through their industry, the colonists gained their property rights.[34] As Locke wrote: "For I ask, whether in the wild woods and uncultivated waste of *America,* left to nature, without any improvement, tillage or husbandry, a thousand acres yield the needy and wretched inhabitants as many conveniences of life, as ten acres of equally fertile land do in *Devonshire,* where they are well cultivated."[35]

In his depiction of the Americas, Locke evokes a landscape in which the agricultural potential is squandered, a description that was integral to his defense of English property rights in the New World. Through labor, the colonists gained rights to dominion over the land. Man transmutes the natural world into a commodity that can be owned. As Locke states: "*As much land* as a man tills, plants, improves, cultivates, and can use the product of, so much is his *property.* He by his labour, does, as it were, inclose it from the common."[36] Locke imagined America as devoid of history, agriculture, and geographic specificity. Famously, Locke wrote: "Thus in the beginning all the world was *America.*"

This evocation of origins—of America as both a wasteland and an untouched surface—resonates with the white sheet of paper.[37] Through labor and industry, the rags of society are transformed into a commodity. In turn, this artificial material of human ingenuity becomes equatable to the surface of the mind—the site for the acquisition of knowledge. And through the act of inscription, the sheet of paper and the mind become the property of the individual. There is not a simple one-to-one relationship between Locke's views on the acquisition of knowledge and property, but there is a movement to see wastelands and blank surfaces, to evoke tabulae rasae and empty spaces for cultivation. This relationship between plowing and writing on an untouched surface, as Bernhard Siegert notes, is not new but rather ancient. The plows that were "used to draw a sacred furrow to demarcate the limits of a new city" also constituted an act of writing. Both the plow and the written indentation "mark the distinction between inside and outside, civilization and barbarism, an inside domain in which the law prevails and one outside in which it does not."[38] Plowing on land and writing on paper infects Locke's own metaphors. In his discussion of memory, he equates the mind of experience to a cornfield, and fleeting ideas become likened to birds and clouds transiting over the field: "*Ideas* in the mind quickly fade, and often vanish quite out of the Understanding, leaving no more footsteps or remaining Characters of themselves, than Shadows do over Fields of Corn."[39] While the pre-sensate mind is a white sheet of paper, the mind of experience is a cornfield.

Yet paper was not a tabula rasa and America was not a wasteland, although they both came to embody similar myths of laboring on an untouched surface. Locke's movement away from the wax tablet as a surface to describe the acquisition of knowledge was not simply a modernization of the ancients but a demarcation of a new vocabulary entrenched in both industry and colonialism. In Locke, a no-place paper emerges, embodying an *idea* of what paper is that persists today in the glowing white sheet of word-processing programs. Integral to the concept of paper-as-mind was a particular form of paper: the white European writing paper that was celebrated as a product of agriculture, industry, and the transformation of waste. On the one hand, paper in Locke is distinct. It is European. It is artificial. It is the transmutation of waste into a commodity for the market. On the other hand, these specific qualities of paper are erased as paper became an abstract idea of a blank surface for writing and for plowing, for imprinting experience, and for demarcating property.

Conclusion

Rembrandt and a New Paper World

The history of print has obscured the importance of paper as an equal force in defining the early modern world. This book, therefore, has focused primarily on works outside of print—from Simone Martini's paintings to legal documents in New Spain. Yet the developing trade in global commodities also transformed paper's ability to convey meaning in the seventeenth century. For the first time since the thirteenth century, Europe began to import paper from a distant geographic area: Asia. At the same time, paper in Europe was achieving an unprecedented abundance.

Pieter Bruegel the Younger (1564–1636) pictorializes the increasing ubiquity of paper in *The Village Lawyer* (1615–22), painted in at least nineteen signed copies (fig. 103).[1] This is not the singular diplomatic letter traveling across the room in Mantegna's Camera degli Sposi (see figs. 53 and 54), in which the display of seigniorial power reveals itself in the precious exchange of epistolary correspondence. Nor is this the careful construction of leather coffers and manuscript bindings at the court of the Duke of Berry to bind, seal, and preserve both the notarial record and the lyrical traces of its poets. Instead, Bruegel revels in the explosion of notarized documents. Sheaths of paper overflow on desks and shelves as they fly away from their folios, only to be stepped on, torn, and forgotten. This is a legal space dominated by the handwritten contract. Bruegel displays an effort to control the paperwork in various burlap sacks, marked for future retrieval. Yet despite the endeavor to file, this is chaos over order, a time defined by paper's profusion.

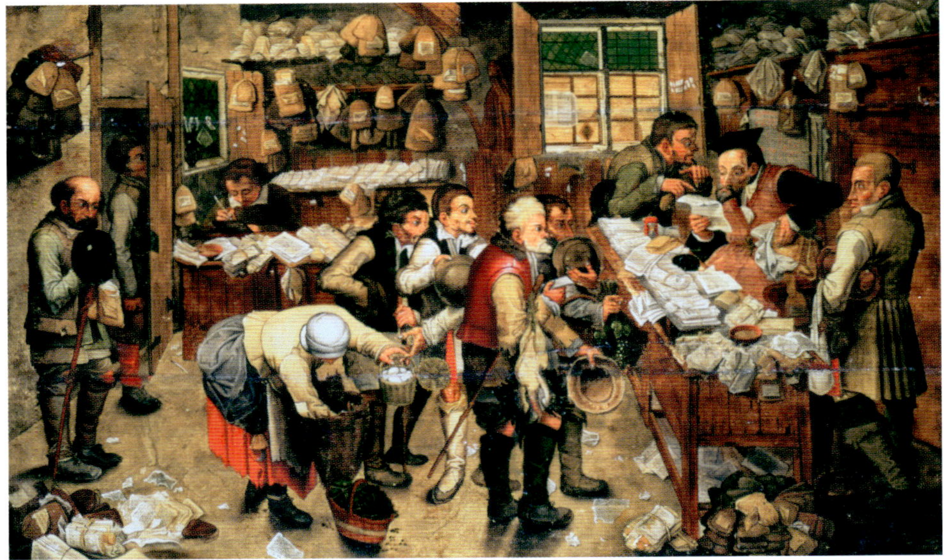

Figure 103. Pieter Bruegel the Younger, *The Village Lawyer*, 1618. Oil on panel, 28¾ × 21⅜ in. (73 × 105 cm). Private collection.

When the Dutch East India Company (Vereenigde Oostindische Compagnie) expanded into Asia, one of the commodities that ships carried back to Amsterdam was paper.[2] Many scholars have noted that this foreign import significantly impacted the late etchings and draftsmanship of Rembrandt van Rijn (1606–1669). These papers have been categorized as "Asian" or "Oriental" because their exact origins remain uncertain. In 2015 a team of paper conservators examined the imported papers employed by Rembrandt and discovered that often these supports were made from *gampi* fibers, and therefore were Japanese.[3] It is unlikely, however, that Rembrandt was aware of the exact origins of his novel paper. Even more, it is uncertain if he knew that this material was made from raw plant material and not textiles. Instead, Rembrandt appreciated the formal qualities that the paper allowed: a lustrous, silky finish and warm, golden tone. The attributes of the paper's ground became integral to his compositions, such as in the *Descent from the Cross*, in which the creamy tan paper evokes the chiaroscuro of nighttime and firelight infusing the scene (fig. 104).[4]

The paper also had meaning for Rembrandt beyond its compositional possibilities: Rembrandt exclusively employed this imported good for the series

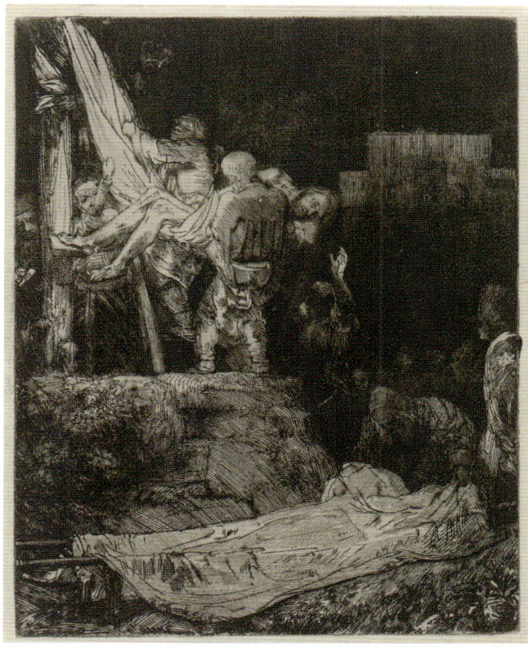

Figure 104. Rembrandt van Rijn, *Descent from the Cross*, 1654. Etching and drypoint on Japanese paper, 8 1/4 × 6 3/8 in. (21 × 16.2 cm). Rijksprentenkabinet, Amsterdam.

of drawings that he executed after Mughal miniatures. In his deployment of the paper as a support for drawings after the Mughal paintings, Rembrandt established a connection between the artistic culture he was emulating and its material presence. Yet this Japanese paper has little in common with the rag-paper tradition in India.[5] For Rembrandt, however, the exact origin point for this paper was lost in the circuits of trade networks. Once it arrived in Amsterdam, it symbolized Eastern origins, conveying material traces of the civilizations Rembrandt sought to evoke in his depictions of biblical antiquity.

Rembrandt was interested in copying Mughal miniatures in part because of an idea that the opaque watercolors offered a view onto an Old Testament antiquity that survived into seventeenth-century India.[6] Or as Leonard Slatkes wrote: "Rembrandt apparently believed that even relatively contemporary Persian and Mughal miniatures transmitted the social customs and dress of biblical days, and he used a variety of these elements to add a note of historical veracity to his work."[7] This seventeenth-century alignment of contemporary religious practice in India with the Old Testament is supported by travel accounts such as the one cited in Chapter 1, *Conformity of the Customs of the Oriental Indians with Those of the Jews and Other Peoples of Antiquity,* in which the author stated that the study of India was a view onto biblical antiquity.

In terms of Rembrandt's artistic practice, scholars cite his drawn copy (fig. 105) after *The Four Mullahs* painting (fig. 106), which inspired his composition *Abraham Entertaining the Angels* (fig. 107).[8] In repurposing the Mughal painting's composition, Rembrandt conceptualized an Old Testament gathering through a contemporary image produced in India. Similar to the confusion around temporality and Eastern antiquities that Simone Martini grappled with in early fourteenth-century Siena, Rembrandt also engaged with seventeenth-century Mughal artistic production as a means to gain representational insight into the Old Testament. For both artists, the particular paper support materialized biblical antiquity.

Yet in his employment of Japanese paper, Rembrandt formed an unexpected hybrid of multiple paper worlds.[9] For his use of these papers deviates from the geography of paper recognized by Dürer when he drafted on carta azzurra in Venice to experiment with specific ideas of facilità and grazia. When Dürer traveled to Venice, he engaged with a commodity employed by Venetian artists to explore specific concepts that defined Venetian artistic practice. Rembrandt neither went to India nor utilized paper from India to consider how the specificity of the paper's ground might contribute to a certain aesthetic discourse. If Rembrandt had worked in the ateliers of the Mughal court, he might have recognized that the predominant paper had more in common with the European support than the Japanese import.

For paper manufactured in India was rag paper. The technology developed under Islamic papermakers in India around the twelfth century, but it did not displace parchment and palm leaves until the fourteenth century (similar to in Europe).[10] In the sixteenth century, paper manufacture took hold with centers established in Kashmir, Delhi, Patna, Shahabad, Kalpi, Ghosunda, Ahmedabad, and Kaghazipura, geographic names that were used to define standard qualities of paper.[11] In turn, paper in India departed from the European tradition because of a greater diversity. The paper world of the Mughal court and its artists was not only indebted to the papermaking technology of the Islamic Empire but also to the imported papers from Asia.[12] The Mughal court, therefore, was at a crossroads of paper histories, a transcultural meeting place embodied in the visual tradition of the album, formed by cutting and pasting into bound books exempla of calligraphy, European prints, and Mughal and Safavid painting (figs. 108, 109). The shared technology of paper was central to the formation of these visual comparisons.

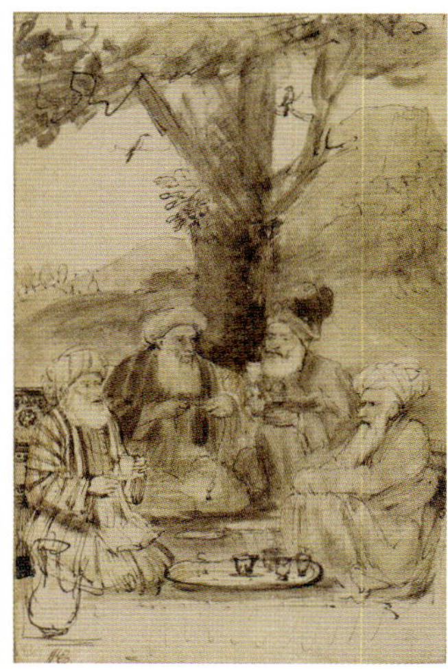

Figure 105. Rembrandt van Rijn, *Four Mughal Elders Sitting under a Tree*, 1656–61. Pen and brown ink on "Oriental paper" prepared with pale brown wash, 7⅝ × 4⅞ in. (19.4 × 12.4 cm). British Museum, London.

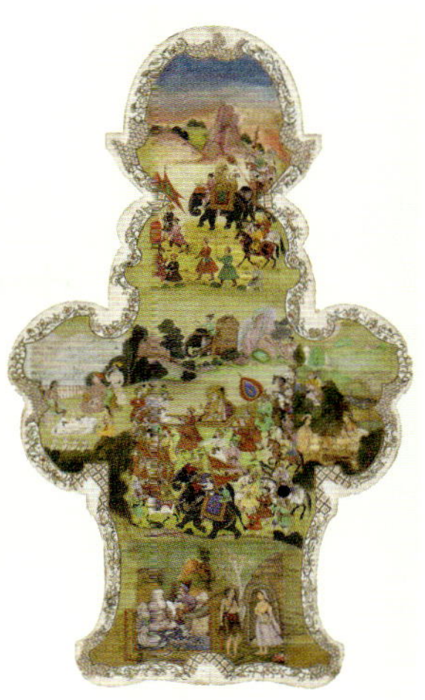

Figure 106. Four Mullahs from panel of paintings in the *Millionenzimmer*, 1627–28. Ink, opaque watercolor, and gold on paper, 36¾ × 21⅞ in. (93.4 × 55.6 cm) (full panel, unframed). Schloss Schönbrunn, Vienna.

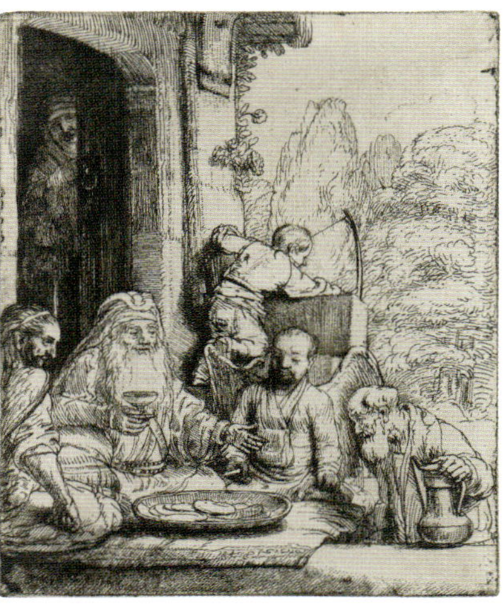

Figure 107. Rembrandt van Rijn, *Abraham Entertaining the Angels*, 1656. Etching and drypoint on paper, 6¼ × 5³⁄₁₆ in. (15.9 × 13.1 cm). Rijksprentenkabinet, Amsterdam.

When Rembrandt copied his Mughal miniatures onto the Japanese paper, he was not tracing a direct line of material influence between the rag-paper tradition of India and Europe, both of which began in the papermaking capitals of the Islamic Empire. Instead, he employed an entirely different tradition of paper. This distinction was not important for Rembrandt, who likely would have categorized both the paper and the Mughal miniatures as "curiosities."[13] Yet Rembrandt's incorporation of papers imported from Asia into his drawings after Mughal miniatures evokes a vague geography of the "East" for a European market. When seen in the context of Bruegel's *Village Lawyer* (see fig. 103), the significance of a rare and imported paper becomes clear in a world dominated by European paper. For all the artists discussed in this book, from Simone Martini to the tlacuilos in New Spain, the particular paper support offered a means to convey a material history that was integral to both the reception of the work and the artist. Yet for Rembrandt, paper was a channel to market and commodify.

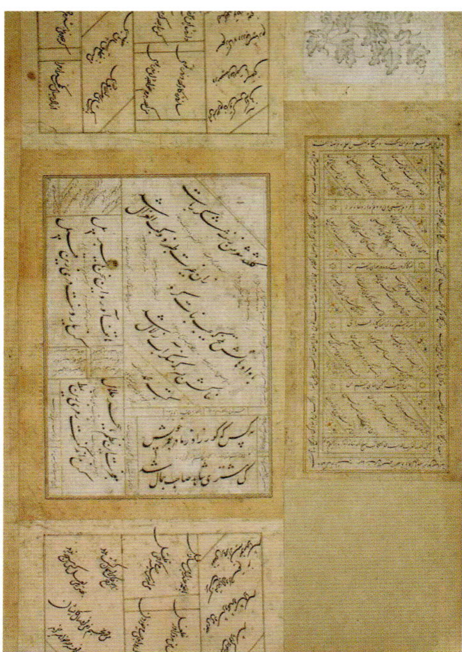

Figure 108. Page of calligraphy in the *Bellini Album*, ca. 1600. Ink, pencil, opaque watercolor, and gold on paper. Metropolitan Museum of Art, New York.

Figure 109. Six devotional scenes in the *Bellini Album*, ca. 1600. Ink, pencil, opaque watercolor, and gold on paper with engravings. Metropolitan Museum of Art, New York.

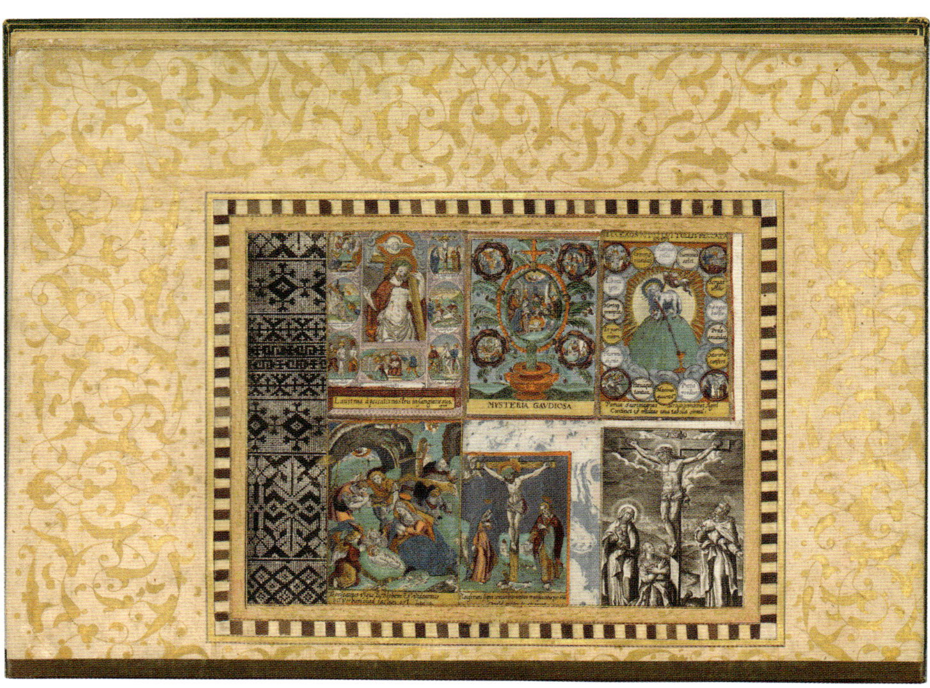

These Eastern papers offered his prints a competitive edge in the paper-print market and became another measure of discernment for the willing connoisseur.

At the end of his career, Rembrandt produced his etchings in a variety of states on a multitude of supports from imported papers to vellum. To take one example, for the first four states of his *Ecce Homo,* thirteen are on Japanese paper (fig. 110), five on a lighter-toned Asian paper frequently referred to as "China paper," and one on vellum. It was not until his fifth state that Rembrandt primarily pulled impressions from Western paper, although there is one known on Japanese paper. The final states of the plate are printed on both Japanese and European paper (figs. 111, 112).[14] These different paper technologies filter through the states to establish difference in the world of print and reproduction.[15] Through the variety of paper, Rembrandt ensured that his prints remained rare, as the increasing proliferation of paper and prints demanded new strategies against the inevitable flow of loss, decay, and forgetting. Yet the mutating states and supports of Rembrandt's late etchings not only reveal an interest in capitalizing on the print market in Amsterdam. They also introduce the image itself as being constantly under a process of metamorphosis. Rembrandt's rendition of the *Ecce Homo* was indebted to Lucas van Leyden (fig. 113), but it was also formed from his own Amsterdam: its town hall, architecture, costume, and historical time. The multiple states and supports of Rembrandt's etchings made evident the necessary conversion that all works undergo to survive. And it is this preservational volatility that defines paper, as its capacity for storage and transmission changes in relationship to other forces, from an increased network of global commodities to our current digital age. This does not harken an end of a paper world, but a transition to a new one in which paper will define a new measure of time and space against other extant media. For the period discussed in this book, paper shifted a previously palimpsestic culture into a site of archives, collections, and bodies of work. Yet as much as paper was a storage material for the historical record, it was defined by a mutability that guaranteed an endless process of transformation.

This is best encapsulated in the twentieth-century writer Anaïs Nin's brief considerations of the ragpicker, at a time when they were fast becoming obsolete. In her reverie "Ragtime" (1938), fragments and "incompleted worlds" form the basis of the ragpicker's living. She describes the textile fragments collected by ragpickers for the paper mills as "the end of objects, and the beginning of transmutation."[16] Following the ragpicker through the city at night, she asks: "Where are all the other things, I say, where are all the things I thought dead?"[17] And in a sort of response, the ragpickers gather together and sing a

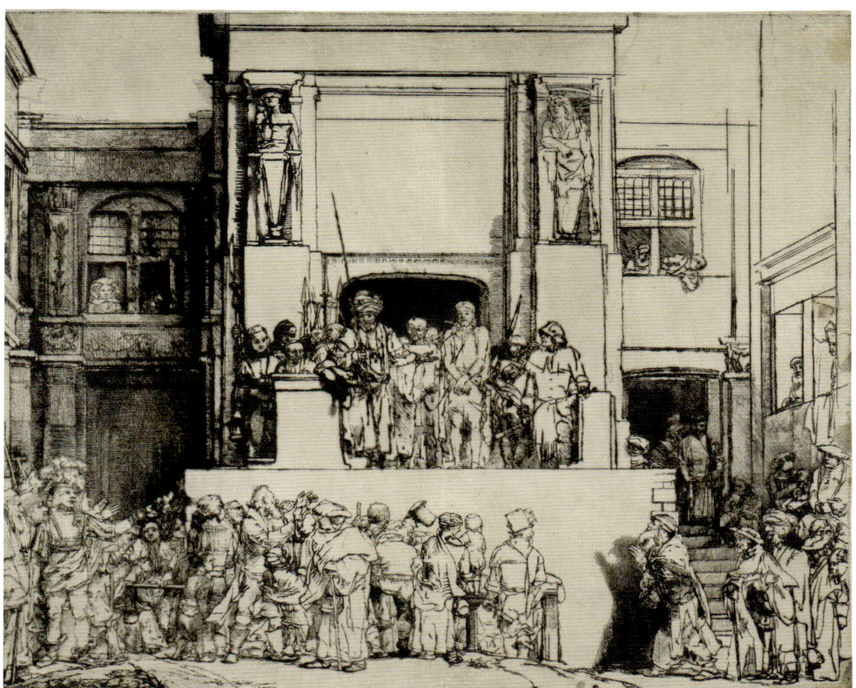

Figure 110. Rembrandt van Rijn, *Ecce Homo*, 1655. Drypoint on Japanese paper (iv), 14 1/16 × 17 1/2 in. (35.8 × 44.4 cm). British Museum, London.

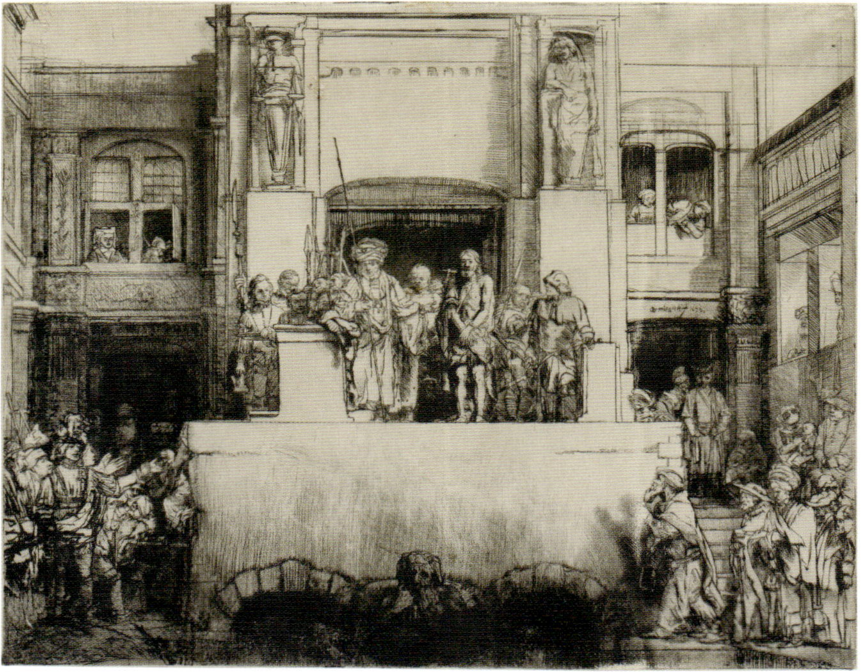

Figure 111. Rembrandt van Rijn, *Ecce Homo*, 1655. Drypoint and pen and ink on Japanese paper (vii/viii), 14 3/16 × 17 3/4 in. (36 × 45 cm). Rijksprentenkabinet, Amsterdam.

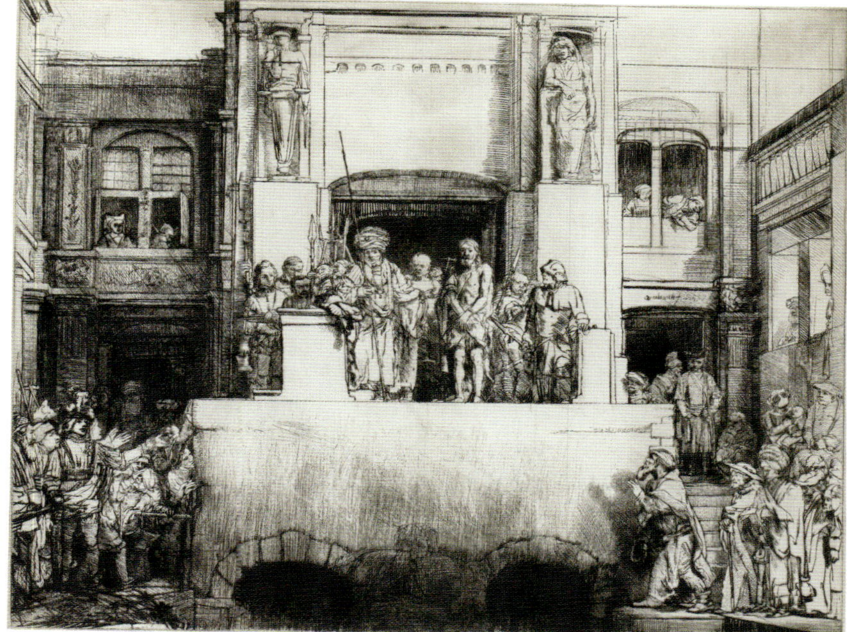

Figure 112. Rembrandt van Rijn, *Ecce Homo*, 1655. Drypoint on European paper (viii/viii), 15 1/16 × 17 15/16 in. (38.3 × 45.5 cm). Rijksprentenkabinet, Amsterdam.

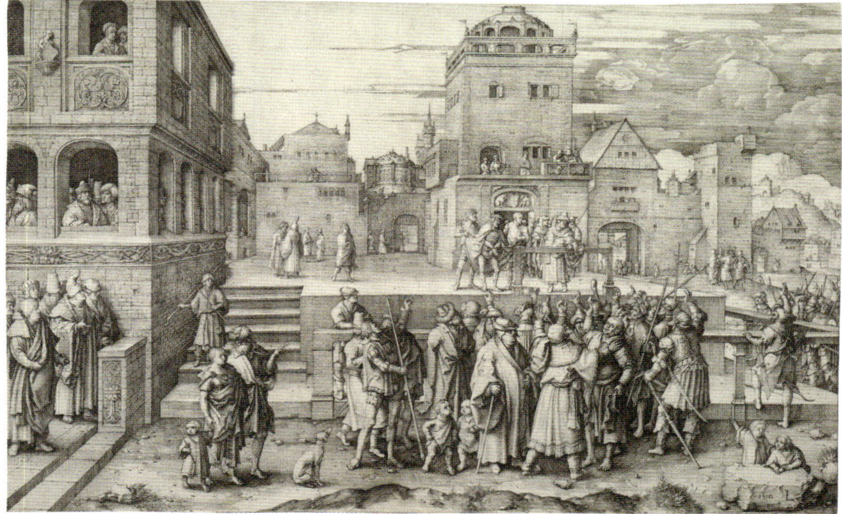

Figure 113. Lucas van Leyden, *Ecce Homo*, 1510. Engraving on paper (i), sheet: 11 3/8 × 17 15/16 in. (28.9 × 45.6 cm). Metropolitan Museum of Art, New York.

ballad, like many early modern ballads about paper and its transmutation from worn and dirty rags to a white surface: "Nothing is lost but it changes / into the new string old string / in the new bag old bag / in the new pan old tin / in the new shoe old leather / in the new silk old hair / in the new hat old straw / in the new man the child / and the new not new / the new not new / the new not new."[18] This nighttime entry into a dreamworld of ragpickers ends with the narrator herself picked up and put in a bag, like all the other pieces of cloth, dresses, ropes, and ribbons collected for the paper mill. In this insertion of herself into the world of rags, Nin suggests that both her own body and the corpus of her written work are no different than the rags that will "not be lost but change" into other substances. She brings the reader into a world of constant transmutation, as nothing is lost but rather everything alters into something else.[19]

In the early modern world, paper was a surface formed from the eradicated traces of textiles transmuted into a material to design a new vision of geography, time, and space. Yet paper itself remains a temporary site for holding dreams, ideas, images, texts, and contracts. It is always caught between supporting the singular archival document and the reproductive proliferation of consumption and waste. But paper, like wax and wood, bronze and papyrus, animal bones and animal skin, is merely a placeholder until a future world and its supports take shape, materializing a new structure for composing history.

NOTES

Introduction

1. Ashcroft, *Albrecht Dürer,* 2:767; Dürer, *Schriftlicher Nachlass,* 2:214–15.

2. In 1525 Easter was on April 16; therefore, Whitsunday was June 4.

3. For example, Hunter, *Old Papermaking;* Hunter, *Primitive Papermaking.*

4. Dürer's *Dream* is in a collected album of prints titled *Kunstbuch Albrechten dürers von Nürnberg,* in Vienna. The album was bound in Mainz in 1560 by Franz Behem. It contains 211 prints and thirteen drawings. The album's provenance has been traced to Dürer's student Hans Döring, who likely presented the album to his patron, Count Reinhard Solms, who gave the album to the archduke. Peter Parshall was the first to consider the actual presentation of Ferdinand's collection as inherent to its meaning and interpretation; see Parshall, "Print Collection of Ferdinand." See also Parshall, "Art and the Theater of Knowledge," 7–36; Anja Grebe, "Albrecht Dürers 'Kunstbücher.'"

5. "Questa carta è oggi appresso di me, tenuta per reliquia . . . " Vasari, *Vita di Michelangelo,* 7.

6. Walsham, "Social History of the Archive," 11. See also Yale, "History of Archives." Walsham's articulation of the archive expands upon the work of Michel Foucault: *The Archaeology of Knowledge* and *The Order of Things;* see also Derrida, *Archive Fever.*

7. Foucault, *Aesthetics, Method, and Epistemology,* 215.

8. On paper, draftsmanship, and writing in the Renaissance, see Rosand, *Drawing Acts;* Barkan, *Michelangelo.*

9. "Come io so che, innanzi che morissi di poco, abruciò nessuno vedessi le fatiche durante da lui e i modi di tentare l'ingegno suo, per non apparire se non perfetto . . . " Vasari, *Vita di Michelangelo,* 1:117.

10. The attention to paper as a material of significance, beyond its ability to testify to authorship and dating, has been sparse in art history. Notable exceptions include Robison, *Paper in Prints;* Krill, *English Artist' Paper;* Fairbanks, *Papermaking and the Art of Watercolour;* Bambach, "Purchases of Cartoon Paper"; Lunning, "Characteristics of Italian Paper"; Ellis, ed., *Historical Perspectives.*

11. On the history of early modern rag paper, see Hunter, *Papermaking;* Barrett, *Papermaking.*

12. Margreta de Grazia and Peter Stallybrass elucidate how making paper—transforming soiled linens into a white surface—informed Elizabethan authors' conception of their work as a process of recycling and renewal; see de Grazia and Stallybrass, "Materiality of the Shakespearean Text," 280. For other studies of paper's matter and how it impacted early modern authors and printers, see Calhoun, "Word Made Flax"; Daybell, *Material Letter in Early Modern England;* Mak, *How the Page Matters.* For paper as evidence in bibliography, see Stevenson, "Briquet and the Future of Paper Studies"; Bidwell, "Study of Paper as Evidence."

13. Maier, *Lusius serius,* 83.

14. Allan Stevenson, *Observations on Paper as Evidence,* 1.

15. For examples of paper costs, see Richardson, *Printing, Writers, and Readers,* 25.

16. Derrida, *Paper Machine*, 43.

17. Febvre and Martin, *Coming of the Book*.

18. Bloom, *Paper before Print*.

19. In *Gutenberg's Europe*, Frédéric Barbier writes: "Increasingly the world became a 'paper world,' to the point, sometimes, of losing its objective immediacy" (39). On the history of late-medieval paper, see Bat-Yehouda, ed., *Papier au Moyen Âge;* Ornato, ed., *Carta occidentale nel tardo Medioevo*.

20. Carter and Dobereiner, "Multispectral Imaging."

21. Bloom, *Paper before Print*, 204. This paper manuscript was originally in the possession of the Florentine cardinal Giovanni Salviati (1490–1553) before entering the Vatican library (Vat. gr. 2200).

22. Weiß, *Zeittafel zur Papiergeschichte*, 28.

23. Ibid., 30.

24. Abbott, "Ninth-Century Fragment."

25. On the making of rag paper in China, see Tsuen-Hsuin, *Chemistry and Chemical Technology;* Tsuen-Hsuin, "Paper and Paper Manuscripts," in *Written on Bamboo and Silk*, 149–74.

26. On papermaking under the Abbasids, see Bloom, *Paper before Print*, 46–89.

27. Rischel, "Analysis of the Papermaker's Choice."

28. See Burns, "Paper Comes to the West."

29. On early Spanish paper, see Hills, "Early Spanish Papers"; Valls i Subirà, *Paper and Watermarks in Catalonia*.

30. For an overview of the current literature on bureaucracy and paperwork, see Kafka, "Paperwork."

31. Bloom, *Paper before Print*, 47–49.

32. Reynolds, "Elder Pliny," 308.

33. In his study of shifting notions of how print transformed the meaning of the devotional and aesthetic object in the Holy Roman Empire under Maximilian I, Christopher Wood asserted that "artists began to leave paper trails marking their own careers." Whereas previously works on parchment were circumscribed to local communities, paper allowed for a new mobility and therefore a "public culture of art." Wood's assertion elucidates the importance of paper, although he primarily approaches his analysis through the structure of print media and the role of the drawing *via* print. By contrast, this book approaches the work on paper through paper itself; see Wood, *Forgery, Replica, Fiction*, 349–50.

34. Innis, *History of Communications*, 15–56. Other prominent works by media theorists include Gitelman, *Paper Knowledge;* Kafka, *Demon of Writing*.

35. On these drawings, see Chapman, *Fra Angelico to Leonardo*, 92–95; Popham and Pouncey, *Italian Drawings*, no. 269; Degenhart and Schmitt, *Corpus der italienischen*, vol. 1, no. 109; Ames-Lewis, "Drawing for Panel-Painting."

36. Also noted by Chapman, *Fra Angelico to Leonardo*, 92–95.

37. In his study of Sienese workshop practice, Hayden Maginnis argued that despite the absence of surviving evidence, paper must have been a crucial tool in the trecento workshop, and he suggests that preparatory drawings emerged with paper. Maginnis, "Craftsman's Genius"; Maginnis, *World of the Early Sienese Painter*, 111–18.

38. Vismann, *Files*, 79.

39. Ibid., 81.

40. Wood, *Forgery, Replica, Fiction*, 30.

Chapter I. Tracking Paper Routes across the Early Modern World

1. The best general overviews of papermaking include Hunter, *Papermaking;* Timothy Barrett, "Early European Papers/Contemporary Conservation Papers," *Paper Conservator* 13 (1989): 1–108.

2. La Lande, *Art de faire le papier*. The translation is Jérôme de La Lande, *The Art of Papermaking* (Kilmurry, Ire.: Ashling, 1976).

3. On the publication, see Cole, *Handicrafts of France;* Dumas and Tresse, "Description des Arts et Métiers"; Huard, "Planches de l'Encyclopédie."

4. La Lande, *Art de faire le papier*, 415.

5. First published as Bloch, "Avènement et conquêtes du Moulin à eau"; Bloch, "Advent and Triumph of the Watermill." For a critique of Bloch's thesis, which does suggest a teleological narrative of Western progress, see Lucas, "Industrial Milling."

6. For an overview, see Hills, "Technical Revolution."

7. Bacon, *New Organon,* 151–52.

8. Ibid., 152.

9. Hills, "Papermaking Stampers"; Reynolds, *Stronger than a Hundred Men,* 82–85.

10. As Richard Hills argues, the developments in European papermaking technology are beholden to the wool industry; Hills, "Technical Revolution," 110.

11. On the late-medieval history of wiredrawing, see Von Stromer, "Innovation und Wachstum."

12. La Lande, *Art of Papermaking,* 37; La Lande, *Art de faire le papier,* 470.

13. Barthes, "Plates of the Encylopedia," 218–35. Originally published in *L'univers de l'Encyclopédie: Les plus belles planches de l'Encylopédie de Diderot et d'Alembert* (Paris: Les Libraires associés, 1964).

14. La Lande notes that the rag-smuggling trade was active in eighteenth-century France, depriving paper mills of their necessary stock. La Lande, *Art de faire le papier,* 506.

15. He remarks that a multitude of plants are used to make paper in China, depending on the province. See ibid., 544.

16. Kaempfer, *History of Japan,* 2:21.

17. Du Halde, *Description de la Chine,* 2:286–301. Translation from Du Halde, *Description of the Empire of China,* 1:367.

18. On the publication of Du Halde's text and its possible encounters with Jesuits, see Wu Huiyi, "Les traductions de François-Xavier Dentrecolles (1664–1741), missionaire en Chine: Localisation et circulation des savoirs," *Etrême-Orient Extrême-Occident* 36 (2013): 49–80. Du Halde's description closely parallels Yingxing's, and Du Halde states that the work on Chinese papermaking "appeared under the present dynasty: It is a curious Collection, and esteemed amongst the Learned." Du Halde, *Description de la Chine,* 367. Although Yingxing published his work in the late Ming Dynasty (1638–44), it is possible that if Du Halde had seen a Jesuit translation of it, he could have accidentally dated it to the Qing Dynasty (1644–1912). For an overview of Song Yingxing's biography and his work, see Schäfer, *Crafting of 10,000 Things.*

19. On the illustrations, see Golas, "'Like Obtaining a Great Treasure,'" 584.

20. Song Yingxing, *T'ien-kung k'ai-wu,* 224. Although by the seventeenth century bamboo paper had become predominant, papermaking in Asia incorporated a vast array of materials, including hemp, flax, ramie, mulberry, paper mulberry, rattan, straw, wheat stalks, sandalwood, seaweed, and the waste of silk cocoons. Rags were used in the earliest manufacture of paper in China, but they quickly were replaced by raw hemp and tree bark. On the variety of materials, see Tsien Tsuen-Hsuin, "Raw Materials for Old Papermaking in China," *Journal of the American Oriental Society* 93, no. 4 (1973): 510–19.

21. Yingxing, *T'ien-kung k'ai wu,* 227.

22. Ibid.

23. Ibid., 228.

24. This text is published in Corsi, "Scholars and Paper-Makers," 86–87.

25. Barrett, "Woman Who Invented Notepaper," 209–10.

26. Salmon, "Mission de Théodose de Lagrené."

27. Although earlier samples of rag paper have been excavated from tombs of the Han Dynasty, paper historians generally trace the invention of paper to Ts'ai Lun, a eunuch at the court of Emperor Ho-Ti in 105 CE. For an overview of the history of papermaking in China, see Tsuen-Hsuin, *Chemistry and Chemical Technology.*

28. Jaucourt, "Papier," 11:846.

29. Ibid.

30. Ibid., 11:856.

31. Prideaux, *Old and New Testament,* 138–41. Houghton Library, for example, has the first edition of the text printed in America in 1815, which pur-

ports to be taken from the sixteenth edition, *The Old and New Testaments Connected in the History of the Jews and Neighboring Nations: 1st American, from the 16th London ed. to Which Is Now Added, the Life of the Author* (Charlestown, Mass.: Published at the Middlesex Bookstore, J. M'kown printer, 1815–16).

32. For an overview of the Bodleian's collection of Arabic manuscripts (and other manuscripts, such as Armenian, Coptic, and Persian) in the seventeenth century, see Wakefield, "Arabic Manuscripts in the Bodleian Library."

33. As Robert Jones points out, war, plunder, and booty were the primary means by which Arabic manuscripts entered into Western collections; see Jones, "Piracy, War, and Acquisition." On the means by which Arabic manuscripts were collected in the early modern period, see also Ghobrial, "Archive of Orientalism."

34. La Créquinière, "De la maniere done les Indiens écrivent, et de ce dont ils se servent au lieu de papier," in *Conformité des coutumes des Indiens orientaux avec celles des Juifs et des autres Peuples de l'Antiquité* (Brussels: G. de Backer, 1703), 186. It was also translated into English: La Créquinière, *The Agreement of the Customs of the East-Indians, with Those of the Jews and Other Ancient People* (London: Printed for W. Davis, at the Black Bull, next the Fleece-Tavern, in Cornhill, 1705).

35. La Créquinière, "De la maniere done les Indiens écrivent," 195.

36. Peter Bower also notes that the Dutch scholar Gerard Meerman (1722–1771) offered a prize for the person who could identify the origins of rag paper; Bower, "White Art." Meerman published his findings in *De Observationes Chartae vulgaris sue linea origine* (Hague Comitum: Nicholaum van Daalen, 1767).

37. It was posited that Greek refugees in Basel invented paper made from linen rags when they could no longer obtain cotton to make paper. Yet the reference to cotton paper, or *carta bombycina,* was a confusion on earlier historians' part, and the manuscripts they are referring to are on linen and hempen rag paper. On the meaning of "carta bombycina" in late-medieval inventories, see Piccard, "Carta bombycina," 46–75. Julius Wiesner was the first to point out that cotton paper was linen rag paper; see Wiesener, "Faijûmer und Uschûmeiner Papiere." See also Jean Irigoin, "Les premiers manuscrits grecs et écrits sur papier et le problème du bombycin," *Scriptorium* 4, no. 2 (1950): 194–204; and A. F. R. Hoernle, "Who Was the Inventor of Rag-Paper?" *Journal of the Royal Asiatic Society* 43 (1903): 663–84.

38. Von Karabacek, "Arabische Papier"; Wiesner, "Faijûmer und Uschmûneier Papiere." For contemporary overviews of paper technology and its use in the book arts of the Middle East, see Loveday, *Islamic Paper;* Jonathan Bloom, *Paper before Print;* Pedersen, "Writing Materials"; Baker, "Note on the Expression"; Baker, "Arab Papermaking"; Déroche, *Islamic Codicology.*

39. Geneviève Humbert articulates many of the questions facing historians of paper in "Manuscript arabe et ses papiers."

40. For translations and discussions of the two texts, see Levey, "Medieval Arabic Bookmaking"; Gacek, "On the Making of Local Paper."

41. Jonathan Bloom, "Papermaking: The Historical Diffusion of an Ancient Technique," in *Mobilities of Knowledge,* https://link.springer.com. This is supported by work on medieval mill technology; see Lucas, *Wind, Water, Work,* 65.

42. Papermakers in Spain did begin to use a zigzag demarcation that some scholars argue was an early form of watermark, although most recently it is thought that the impression of the zigzag aided in binding papers together. On the zigzag, see Estève, "Zigzag dans les papiers arabes," 40–49.

43. See Irigoin, "Types des formes"; Malachi Beit-Arié, "Quantitative Typology of Oriental Paper Patterns," in Bat-Yehouda, ed., *Papier au Moyen Âge,* 41–53; Humbert, "Papiers non filigranés."

44. Voorn, "Brief History of the Sizing of Paper."

45. Two primary examples that Briquet mentions are Casiri, *Bibliotheca Arabico-Hispana Escurialensis;* and Assemanus, *Bibliotheca orientalis Clementi-*

no-vaticana. Briquet, "Recherches sur les premiers papiers," 23–25.

46. Ibid.

47. Chambers, "Condottiere and His Books."

48. Grabar, *Illustrations of the Maqamat.*

49. Nelson, "Slide Lecture"; Bohrer, "Photographic Perspectives."

50. This trend began with Frederick Starr, an ethnologist who also published work on papermaking in Mexico and attempted to understand the variety of Mesoamerican papers through the surviving twentieth-century papermaking practices of the Otomí; for example, see Starr, *In Indian Mexico;* Sandstrom and Sandstrom, *Traditional Papermaking.*

51. Von Hagen, *Aztec and Maya Papermakers,* 9.

52. Hunter, *Primitive Papermaking,* 14.

53. Diego de Landa, *Relación de las cosas de Yucatán,* 33; *Landa's Relación de las cosas de Yucatan,* 28. On Mesoamerican histories of the book, see Boone, *Stories in Red and Black;* Schroeder, "Writing Two Cultures"; Brotherston, *Painted Books from Mexico.* Also see Robertson, *Mexican Manuscript Painting.*

54. D'Anghera, *De Orbe Novo,* 2:41.

55. Stallybrass, Chartier, Mowery, and Wolfe, "Hamlet's Tables and Technologies." On erasure in Maya and Mixtec manuscripts, see Carter and Dobereiner, "Multispectral Imaging," 722–23. Smith, "Why the Second Codex Selden Was Painted." The Codex Selden, a well-known palimpsestic manuscript, is on animal hide and not paper.

56. Douglas, *In the Palace of Nezahualcoyotl,* 41–43.

57. Hunter, *Primitive Papermaking,* 15–16.

58. Ibid., 14.

59. Berdan and Durand-Forest, *Matrícula de Tributos;* Berdan, "Imperial Tribute Roll"; Batalla Rosado, "Scribes Who Painted."

60. See Bleichmar, "History in Pictures."

61. Purchas, *Purchas His Pilgrimes,* 3:1084.

62. Pliny, *Natural History,* trans. H. Rackham (Cambridge, Mass.: Harvard University Press, 1938), 13:24.

63. Sahagún, *General History of the Things of New Spain,* pt. 3, 69.

64. Ibid. On the ritual use of amatl, see Lenz, *Papel indigena mexicano.*

65. On the processes of destruction and renewal as structuring the Aztec philosophical landscape, see Maffie, *Aztec Philosophy.*

66. Arnold, "Paper Rituals," 229. See also Arnold, "Paper Ties to Land."

67. For overviews on Arabic printing in the West, see Jones, "Medici Oriental Press in Rome"; Lincoln, "Gospel Lessons"; Dannenfeldt, "Renaissance Humanists"; Pedersen, "Printed Books," in *The Arabic Book,* ed. Robert Hillenbrand, trans. Geoffrey French (Princeton: Princeton University Press, 1984), 131–42; Roper, "Printed in Europe."

Chapter II. Forgetting Paper's Origins

1. Peter the Venerable, *Against the Inveterate Obduracy of the Jews,* 218.

2. Blum, *On the Origin of Paper,* 26–27.

3. Abu-Lughod, *Before European Hegemony.* See also Nagel, *Some Discoveries of* 1492.

4. Abu-Lughod, *Before European Hegemony,* 67.

5. Goitein, *Mediterranean Society.*

6. Ibid., 1:81. Medieval paper made in the capitals of Baghdad, Damascus, and Cairo does not have the defining feature of the watermark that scholars use to situate European paper. Therefore, paper historians often identify stocks according to the imprints of the chain and laid lines left by the paper mold. Yet there are also surviving manuscripts made with wireless paper molds, meaning that there are neither laid nor chain lines; see Humbert, "Papiers non filigranés utilisés"; Beit-Arié, "Oriental Arabic Paper"; Burns, "Paper Revolution in Europe," 5.

7. On the paper documents in the archives of James I of Aragon, see Plano, "Inventario analítico de

documentos." Cabañas and Trenchs, "Registros de cancilleria de la coronoa de Aragón"; Burns, "Paper Revolution in Europe"; Burns, *Society and Documentation in Crusader Valencia*. For a general history of paper in Spain, see Valls I Subirà, *Historia del papel en Espagña*. For a history of paper in artistic practice in Spain, see McTigue, "Paper and Papermaking in Spain."

8. Burns, "Paper Revolution in Europe," 9. Also on medieval Spanish paper, see Valls i Subirà, *Paper and Watermarks in Catalonia*; Vass i Subirà, *Historia del papel en España*.

9. Burns, *Society and Documentation*, 159–60.

10. Cited in Piccard, "Carta bombycina," 55.

11. The document was first published by Briquet, *Papiers et filigranes*, fol. 304. It was discussed in Vass i Subirà, *Historia del papel en Espana*, 1:170; Guichard, "Du parchemin au papier."

12. Amari, *I diplomi arabi*, 239–40, 451, 285–87, 468.

13. Ibid., 78–79, 412.

14. Ibid., 115–18, 419.

15. Williams, "Unfolding Diplomatic Paper," 504. On note-taking and paper, see Heesen, "Notebook"; Blair, "Note-Taking"; Soll, "From Note-Taking to Data Banks."

16. See Maire-Vigueur, "Révolution documentaire"; Cammarosano, *Italia medievale*. As many scholars have argued, the rise of paper in Italy was connected to the revival of Roman law, as the notary became a necessary profession. Bloom, *Paper before Print*, 210; Burns, *Society and Documentation*, 15.

17. Cossar, *Clerical Households*, 35.

18. Ibid., 25.

19. Waley, *Siena and the Sienese*, 156.

20. Prazniak, "Siena on the Silk Roads."

21. Continelli, *L'Archivio dell'Ufficcio dei Memoriali*. Also see Wray, *Communities and Crisis*, 11–56. For an overview of the current methodological approaches to early modern archives and record-keeping, see Walsham, "Social History of the Archive," 9–48.

22. As Markus Friedrich writes: "From the late Middle Ages onward, this habit of 'thinking with archives' began to influence an ever-growing set of social practices. Archives came to permeate economic and legal life; archives started to influence religious life—not only did the administration of the churches become bureaucratic, but also individual believers now had to keep and preserve (that is: to archive) at least a few relevant documents." Friedrich, "Archival Practices," 469. Archives provide regularity to the accumulation so that they may be consulted in the future to establish laws, precedent, and history. In contrast, the Cairo Genizah is not strictly considered an archive, because the documents lack a system of registration and cannot be easily accessed by future members of the community. Ketelaar, "Records Out and Archives In," 205.

23. Gasparinetti, "Ein altes Statut von Bologna," 45–47.

24. Lugli, "Hidden in Plain Sight."

25. For a comprehensive account of the freco's importance in its political context, see Norman, *Siena and the Virgin*.

26. On the paper in the fresco, see Muller, "Paper in Simone Martini's Frescoed *Maestà*." The material was first identified as parchment; see Borsook, *Mural Painters of Tuscany*, 21. See also Bagnoli, "I tempi della Maestà," 116.

27. Waley, *Siena and the Sienese*, 156.

28. The presence of Mamluk metalware, *tiraz* textiles, and other luxury materials from the Holy Land in thirteenth-century Sienese painting illustrates extensive trade with the Mongol Empire, an economy in luxury goods that was most likely transacted on non-European paper—at least from the merchants in North Africa, Iraq, and Iran. On the trade, see Abulafia, "Role of Trade in Muslim-Christian Contact."

29. Mack and Zakariya, "Pseudo-Arabic."

30. Thank you to Alexander Nagel for pointing this out to me.

31. Jerome, "Prologue to Job," in Jerome, *Letters and Select Work*, 1037–38; Jerome, *Biblia sacra*, 1:731.

32. Jerome, "Letter to Pope Damasus: Beginning of the Preface for the Gospels," in *Letters and Select*

Works, 1030; Jerome, *Biblia sacra,* 1:3. For an overview of Jerome's understanding of translation within a longue durée of translation theory, see Venuti, "Genealogies of Translation Theory."

33. Mack, *Bazaar to Piazza,* 71.

34. Nagel, *Some Discoveries of* 1492, 16.

35. Mack and Zakariya, "Pseudo-Arabic," 168.

36. In her work on the Islamic influence on Venetian architecture, Deborah Howard considers the importation of Arabic books an important visual source for Venetian artisans, and she cites the inventories of Venetian merchants who died in Damascus and listed books among their possessions. She notes that inventories distinguish between papers: *bona carta* and *carta bambaxina.* Howard, *Venice and the East,* 59.

37. The first inventory of Arabic manuscripts in the Vatican was made in 1481. Most of these manuscripts came into the possession of the Vatican from the Councils of Ferrara in 1438 and Florence in 1441. The inventory describes all of the works as *ex papyro,* presumably referring to paper and not papyrus. On the earliest Arabic manuscripts in the Vatican collections, see Vida, *Richerche.*

38. Nagel, *Some Discoveries of* 1492, 8.

39. On Mantegna and pseudo-scripts, see Nagel, "Twenty-Five Notes."

40. Antenhofer, "Gonzaga und Mantua," 43–44.

41. Nagel, "Twenty-Five Notes on Pseudoscript," 244. Lipshitz, "Performance as Profanation."

42. Thomas A. Fudge, "Jan Hus at Calvary: The Text of an Early Fifteenth-Century *Passio,*" *Journal of Moravian History* 11 (2011): 45–81.

43. The text is known as the *Passio etc. secundum Iohannem Barbatum, rusticum quadratum,* and is considered an eyewitness account of Hus's execution, written between 1415 and 1418. The Latin text is published in Václav Novotný, ed., *Fontes rerum bohemicarum* (Prague: Nákladem nadáni Františka Palackého, 1932): 8:14–24. The translation is from Fudge, "Jan Hus at Calvary," 68.

44. On these illustrations, see Karl Küp, "The Illustrations for Ulrich von Richenthal's Chronicle of the Council of Constance in Manuscripts and Books," *Papers of the Bibliographic Society of America* 34, no. 1 (1940): 1–16.

45. Ferrari, "Gonzaga Archives of Mantua."

Chapter III. The Model of Loss in Late-Medieval Drawing

1. Guillaume de Machaut, *Le livre dou voir dit/The Book of the True Poem,* ed. Daniel Leech-Wilkinson, trans. R. Barton Palmer (New York: Garland, 1998), 3145.

2. On the role of images as physical impressions into the psyche in Machaut, see Perkinson, "Rethinking the Origins of Portraiture."

3. The scholar Sylvia Huot cites pivotal factors for this shift, including "the analogies between scribal and poetic practice, the consciousness of the vernacular lyric or lyrico-narrative poet as a writer, the attention to the act of poetic composition and the primacy of composition over performance." Huot, *From Song to Book,* 234–35.

4. On Machaut's attention to his books as bound physical objects, see Brownlee, *Poetic Identity,* 15; McGrady, *Controlling Readers.* McGrady also attends to the ways in which Machaut's readers were formative in constructing his physical legacy in bound works. McGrady, "Machaut and His Material Legacy"; Williams, "Author's Role."

5. Froissart, *Prison amoureuse,* 42–43.

6. Ibid., 210–11.

7. Jean Froissart, "Le Dit dou Florin/The Tale of the Florin," in *An Anthology of Narrative and Lyric Poetry,* ed. and trans. Kristen M. Figg and R. Barton Palmer (New York: Routledge, 2001), 498–99.

8. Jean Froissart, "Le joli buisson de jonece/The Bush of Youth," in Figg and Palmer, eds., *Anthology of Narrative and Lyric Poetry,* 268–69.

9. One scholar notes that Machaut's concern for the book "where he put all his things" was "a logical ex-

tension of the clerkly cast of literature in this period." Leach, *Guillaume de Machuat*, 1–2.

10. Cerquiglini-Toulet, *Color of Melancholy*, 40. First published as *La couleur de la mélancolie: La frequentation des livres aux XIVe siècle* (Paris: Hatier, 1993).

11. On Froissart and his concern with transmission, see two works by Peter F. Ainsworth: "Patterns of Transmission in the *Chroniques*"; and *Jean Froissart and the Fabric of History*.

12. Froissart, *Prison amoureuse*, 244–46.

13. Jacqueline Cerquiglini-Toulet illustrates that the fourteenth century was a time of "literary reflexivity—of taking the very act of writing as the object of writing." Cerquiglini-Toulet, *Color of Melancholy*, 53.

14. On Jean de Berry's patronage of Machaut, see Earp, "Machaut's Role," 489–90.

15. Roger Chartier suggests that Machaut and Froissart became aware of the materials of their craft as the interrelationships among book, work, and author shifted in the fourteenth century; see Chartier, "Foreword," in *The Color of Melancholy: The Use of Books in the Fourteenth Century*, trans. Lydia G. Cochrane (Baltimore: Johns Hopkins University Press, 1997), xiv.

16. For the canonical work on the importance of voice in medieval lyric, see Zumthor, *Introduction à la poésie orale*. For an important counter to Zumthor's argument, illustrating the relevance of writing to lyric, see Kendrick, *Game of Love*.

17. On Aristotelian theories of voice and troubadour lyric, see Nichols, "Voice and Writing."

18. For examples of these amorphous works, see Holcomb, *Pen and Parchment*.

19. Eberhard König, "How did Illuminators Draw?" *Master Drawings* 41, no. 3 (2003): 218.

20. Ainsworth, "Review," 305.

21. Buchthal, *Musterbuch of Wolfenbüttel*, 13.

22. Geymonat, "Drawing, Memory and Imagination," 221.

23. Ibid., 284.

24. Hans Belting, "Zwischen Gotik und Byzanz: Gedanken zur Geschichte der sächsischen Buchmalerei im 13. Jahrhundert," *Zeitschrift für Kunstgeschichte* 41 (1978): 217–57; Buchthal, *Musterbuch of Wolfenbüttel*, 64–65.

25. On the material history of the work, see Milde, "Zum Wolfenbütteler Musterbuch."

26. Geymonat, "Drawing, Memory, and Imagination," 220–85.

27. Scholars such as Ulrike Jenni have pointed out the complicated ways in which works grouped under the term "model book" are more complicated than merely providing a pattern book for the workshop; see Jenni, "Vom mittelaterlichen Musterbuch," 139–41; Jenni, "Phenomena of Change." On the terminology of model books, see Scheller, *Exemplum*, 54–59.

28. As Jonathan Alexander notes, the texts copied in manuscript scriptoria would have been "largely dictated by the abbot and the officials under him, the prior, sacrist, or librarian, for example." Alexander, "Facsimiles, Copies, and Variations," 61.

29. On the iconographic tradition of evangelist portraits, see Friend, "Portraits of the Evangelists, Pt. I"; Friend, "Portraits of the Evangelists, Pt. II."

30. On the tradition of evangelist iconography and direct or indirect inspiration, see Nelson, "Inspired or Inspiring Evangelist."

31. For the literature on this drawing book, see Scheller, *Exemplum*, 233–40; Kreuter-Eggemann, *Skizzenbuch des "Jaques Daliwe"*; Jenni, *Skizzenbuch des Jaques Daliwe*; Köllner, "Skizzenbuch des 'Jaques Daliwe'"; Frinta, "Reviewed Work(s)"; Scheller, "Reviewed Work(s)." While Daliwe was the owner of the work or the maker, the mastery of the script in which the name is inscribed suggests the training of a scriptorium hand. As Scheller states, it is a "professional" script; Scheller, "Reviewed Work(s)," 206.

32. Frinta suggests that Daliwe's work on *The Flagellation* was the model for the Limbourg brothers, although given the strange innovations in Daliwe's composition, it seems more likely that he was the one playing around with a previous model; Frinta, "Reviewed Work(s)," 100. Jenni, *Skizzenbuch des Jaques Daliwe*, 30–31. Scheller notes a marked difference

between the courtly style of the Limbourg brothers and the emergent Netherlandish "realism" of Daliwe's work, arguing that perhaps he did not work in the shop but was merely familiar with the patrons; Scheller, "Reviewed Work(s)," 207.

33. This is the 1402 inventory, which cites a manuscript (generally thought to be the Brussels Hours, or *Très-belles heures de Jehan de France, Duc de Berry*) "enluminées et ystoriées de la main de Jacquemart de Odin." There is then a 1412 inventory, which cites the *Grandes heures du Duc de Berry* as "Jacquemart de Hodin et autres ouvriers de Monseigneur." For an overview of the literature on Jacquemart de Hesdin, see König, *Vom Psalter zum Stundenbuch*.

34. On the development of this iconography in the Franco-Burgundian context, see Wixom, "Enthroned Madonna with the Writing Christ Child."

35. Parkhurst, "Madonna of the Writing Christ Child," 306. See also Minnis, *Medieval Theory of Authorship*, 39.

36. Froissart, *Chroniques* (2001–ca. 2004), 1:77.

37. Bratu, "'Je, acteur de ce livre,'" 201.

38. Guillaume de Machaut, "Le dit de la fleur de lis et de la marguerite," in Fourrier, ed., *"Dits" et "debats,"* 290.

39. Huot, *From Song to Book*, 322.

40. Scholars suggest that Christ writing on a scroll, while held in the lap of his mother, alludes to his embodiment of the Word from the opening lines of John: "In the beginning was the word, and the word was with God, and the word was God" (John 1:1). Laguna-Chevillotte, "Christ-Verbe," 409.

41. There is greater stylistic disparity in Morgan's book than in the Berlin book, but scholars' ability to disentangle the hands is also muffled by the inclusion of at least one other draftsman's hand (apart from the primary two in the style of Pucelle and Jacquemart), who clumsily outlined and included his own models and faces within the book.

42. Meiss, *French Painting*, 249. In turn, Panofsky remarks on Jacquemart's explicit citation of Pucelle's *Heures of Jeanne de Navarre* in his own *Petites heures*. Panofsky, *Early Netherlandish Painting*, 43.

43. Traditionally, medieval draftsmanship has been in the context of the *Gedankenbild* and the exempla, in which artists established their authority by participating in a chain of models that did not trace to nature but to older, revered artworks; see Von Schlosser, "Vademecum eines fahrenden Gesellen"; Scheller, *Exemplum*, 1995.

44. Fry, "On a Fourteenth-Century Sketchbook," 32. Froissart discusses the savage ball in the fourth book of his *Chroniques*, in a chapter titled "Le bal des ardents."

45. Their portraits are likened to the statues of the king and queen in Poitiers; see Châtelet, "Artiste à la cour de Charles VI," 119. Stephen Perkinson has discussed the importance of the French court circa 1400 for expressing a "nascent interest" in portraiture; Perkinson, "Rethinking the Origins of Portraiture," 135.

46. Froissart, *Chroniques* (1997), 2:336.

47. Camille, *Medieval Art of Love*, 32.

48. Voelke, "Two New Drawings," 245. It also is possible that both the first and second groups were made by the same artist over a period of shifting stylistic change and technical mastery. As one scholar notes: "In any case it seems likely that the best drawings, those at the beginning of the volume, were all executed over earlier ones that had been erased." Jenni, "Vom mittelaterlichen Musterbuch," 142.

49. Rouse and Rouse, "Vocabulary of Wax Tablets," 221.

50. Ibid., 220.

51. Alain of Lille, *Plaint of Nature*, prose 2:3–8, 108. Alain de Lille, *De Planctu Naturae*, in *Studi medievali*, ed. N. M. Häring, 3, no. 19 (1978): 797–879.

52. Bourlet, "Tabletiers parisiens."

53. For the most recent literature on this work, see König and Theisen, *Wiener musterbuches;* Scheller, *Exemplum*, 226–32.

54. Cennini, *Cennino Cennini's "Il Libro dell'arte,"* 35.

55. For the history of boxwood tablets in antiquity,

see Roman, "History of Lost Tablets"; Propertius, *Elegies*, ed. and trans. G. P. Goold (Cambridge, Mass.: Harvard University Press, 1990), 3:23:21–22.

56. Propertius, *Elegies*, 3:23:8–10.

57. Cennini, *Cennino Cennini's "Il libro dell'arte,"* 35.

58. Meder also argues that the paper replaced the wax ground of the workshop models; Meder, *Mastery of Drawing*, 136.

59. Froissart, *Prison amoureuse*, Ll. 1248–52.

60. Froissart, *Prison amoureuse*, 2114–18.

61. Cerquiglini-Toulet, *Color of Melancholy*, 123. Stephen Perkinson also points out how the *coffret* in Froissart is a metaphor for memory; see Stephen Perkinson, *The Likeness of the King: A Prehistory of Portraiture in Late Medieval France* (Chicago: University of Chicago Press, 2009), 159–65.

62. Ainsworth, "Patterns of Transmission," 23.

63. Cerquiglini-Toulet, *Color of Melancholy*, 226.

Chapter IV. Albrecht Dürer and the Geography of Paper

1. Ashcroft, *Albrecht Dürer*, 1:52; Buck, "Tradition als Herausforderung"; Kemperdick, *Martin Schongauer*, 18–19, 69–70, 276–77; Smith, "Albrecht Dürer as Collector," 8.

2. For the most important recent work on Dürer's acute awareness of reception, see Shira Brisman, *Albrecht Dürer and the Epistolary Mode of Address* (Chicago: University of Chicago Press, 2016).

3. On these drawings, see Wood, *Forgery, Replica, and Fiction*, 348. See also Smith, "Albrecht Dürer as Collector," 11.

4. Wood, *Forgery, Replica, and Fiction*, 347–48.

5. For the artistic interchange between Dürer and Venetian artists, see Smith,

"'Germania' and 'Italia'"; Koreny, "Venice and Dürer"; Morrall, "Dürer and Venice." On Dürer's blue-paper drawings, see Strauss, *Complete Drawings*, 2:914–58; Winkler, *Zeichnungen Albrecht Dürers*, 2:89–90; Schröder and Sternath, *Albrecht Dürer*, 226–57.

6. Anzelewsky, *Albrecht Dürer*, cat. 93, 187–99; see also Humfrey, "Dürer's *Feast of the Rose Garlands*," 22; Kotková, ed., *Albrecht Dürer*.

7. Dürer, *Schriftlicher Nachlass*, 1:57. Dürer makes this statement in a letter to Pirckheimer; see Ashcroft, *Albrecht Dürer*, 1:163–65. Scholars have always interpreted this statement to refer to *Jesus among the Doctors*, although, as Ashcroft points out, "*ein ander quar*" may also signify a frame that Dürer designed in the Venetian style for *Feast of the Rose Garlands* (ibid., 1:165n3).

8. Brückle, "Historical Manufacture of Blue-Coloured Paper," 20. Also on blue paper, see Brückle, "Blue-Colored Paper in Drawings."

9. Brückle also suggests that the Venetian preference for the commodity may be linked to its Eastern trade. See Brückle, "Historical Manufacture of Blue-Coloured Paper," 20.

10. On the importance of prepared grounds in various hues, see de Tolnay, *History and Technique*, 74; Ames-Lewis, *Drawing in Early Renaissance Italy*, 35–45; Landau and Parshall, *Renaissance Print*, 183–84.

11. Whistler, *Venice and Drawing*, 109–10.

12. As Katherine Luber notes, before Dürer's Venetian journey in 1505–7, Dürer did not complete any prototype for preparatory work, with the exception of *Virgin with a Multitude of Animals;* Luber, *Albrecht Dürer and the Venetian Renaissance*, 81. Erwin Panofsky maintained that the drawings made in Venice should be seen as "equivalents, rather than anticipations, of paintings"; Panofsky, *Life and Art of Albrecht Dürer*, 118. Nevertheless, Friedrich Winkler maintains that they are preparatory studies; Winkler, *Zeichnungen Albrecht Dürers*, 2:408. On these drawings, see also Röver-Kann, *Dürer-Zeit*, cats. 27, 98.

13. Panofsky, *Life and Art of Albrecht Dürer*, 118.

14. Schröder and Sternath, *Albrecht Dürer*, cats. 102, 103; 334. See also Robison and Schröder, eds.,

Albrecht Dürer, cats. 50, 51; 164–65. The reasons for dividing the sheet remain unclear, yet financial considerations at sale likely played a part; see Schröder and Sternath, *Albrecht Dürer,* 328. On the importance of seeing these two studies as a unified work, see also Robison and Schröder, eds., *Albrecht Dürer,* cats. 48, 49; 157.

15. Robison and Schröder, eds., *Albrecht Dürer,* cats. 48, 49; 157.

16. Strauss suggested that most of Dürer's studies for his woodcuts and engravings were ruined in the process of tracing and needling; see Strauss, *Complete Drawings,* v.

17. Whistler, *Venice and Drawing,* 109–10.

18. Ashcroft, *Albrecht Dürer,* 1:386; Dürer, *Schriftlicher Nachlass,* 2:393–94.

19. Ashcroft, *Albrecht Dürer,* 1:160; Dürer, *Schriftlicher Nachlass,* 1:55.

20. Aschcroft, *Albrecht Dürer,* 1:175. As many scholars have noted, the painting seems to have actually taken Dürer longer than five months.

21. Ashcroft, *Albrecht Dürer,* 1:159; Dürer, *Schriftlicher Nachlass,* 1:55.

22. Katherine Luber suggests that Dürer's use of the term "*erhaben*" (translated as "sublime") is best realized in the Italian term "*grazia.*" For the complications of the term and its contradictions, see Emison, "Grazia."

23. See Gruber and Müller, "Künstler und Gesellschaft"; Clemens, "Michelangelo on Effort."

24. Also cited in Clemens, "Michelangelo on Effort," 302.

25. Pliny the Elder, *Naturalis historia,* 35.40.

26. Ashcroft, *Albrecht Dürer,* 1:176.

27. Cennini, *Cennino Cennini's 'Il libro dell'arte,* 37.

28. Panofsky, *Life and Art of Albrecht Dürer,* 134.

29. Ibid.

30. On the drawings for the Heller Altarpiece, see Strauss, *Complete Drawings,* 2:1016–52; Winkler, *Zeichnungen Albrecht Dürers,* 121–23; Schröder and Sternath, *Albrecht Dürer,* 358–71.

31. Wölfflin, *Drawings of Albrecht Dürer,* 10.

32. Jochen Sander and Johann Schulz, "'I Will Make Something That Not Many Men Can Equal': Dürer and the *Heller Altarpiece,*" in *Albrecht Dürer: His Art in Context* (Munich: Prestel, 2013), 221.

33. The original central panel by Dürer was destroyed in a fire in 1729, and the work is now known through a copy in Frankfurt. For a history of the altarpiece, see ibid., 219–23.

34. Strauss defines a certain category of Dürer drawings as a "file," like the sketchbooks of Pisanello and Bellini; see Strauss, *Complete Drawings,* 1:v. Tietze calls these works "*Detailstudien*"; see Tietze, *Dürer als Zeichner und Aquarelist,* 24–27. For an overview of the variety of Dürer's drawing practices, see Christiane Andersson and Larry Silver, "Dürer's Drawings," in Silver and Smith, eds., *Essential Dürer,* 12–34; Andrew Robison, "The Drawings of Albrecht Dürer," in Robison and Schröder, eds., *Albrecht Dürer,* 17–43.

35. In just one example of their contemporary iconic status, the pop singer Justin Bieber tattooed Dürer's hands onto his calf.

36. Ashcroft, *Albrecht Dürer,* 1:210.

37. Ibid., 1:211.

38. Ibid., 1:220; Dürer, *Schriftlicher Nachlass,* 1:69.

39. Ashcroft, *Albrecht Dürer,* 1:221; Dürer, *Schriftlicher Nachlass,* 1:70.

40. Ashcroft, *Albrecht Dürer,* 1:217; Dürer, *Schriftlicher Nachlass,* 1:67.

41. Ashcroft, *Albrecht Dürer,* 1:219.

42. Ibid., 1:214; Dürer, *Schriftlicher Nachlass,* 1:66.

43. Ashcroft, *Albrecht Dürer,* 1:214; Dürer, *Schriftlicher Nachlass,* 1:66.

44. Ashcroft, *Albrecht Dürer,* 1:47, 224; Dürer, *Schriftlicher Nachlass,* 1:72.

45. Tietze, *Dürer als Zeichner und Aquarellist,* 5.

Chapter V. Paper

1. Thank you to Donald Farnsworth for bringing this watermark to my attention.

2. Campbell, "Finding Folds."

3. Halevi, "Christian Impurity," 917–45.

4. There has been significant scholarship attending the complexity of coexsting systems of writing and literacy in New Spain; see Boone and Mignolo, eds., *Writing without Words;* Rappaport and Cummins, *Beyond the Lettered City.*

5. Folio 19r is part of a Mesoamerican manuscript known as a *tonalmatl,* or "book of days," a divinatory almanac recording the 260-day cycle. This tonalmatl was made for a European audience on European paper, and folio 19r contains multiple registers of language. There is the image by the indigenous scribe and the hand of an annotator, who provides the name and attributes of the deity. A second hand glosses the first hand. In turn, a third hand, in a scrawling cursive, builds on the commentary of the first hand. This tonamatl is part of the Codex Telleriano-Remensis, which also contains two other manuscripts annotated by the same scribes and artists. The almanac is preceded by a *veintena,* or a depiction of the annual eighteen rituals that demarcated the Mesoamerican calendar, and it is followed by a *xiuhamatl,* or "book of years," which recounts a pictorial record on Aztec history from the twelfth to the sixteenth century; see Keber, *Codex Telleriano-Remensis,* 123.

6. Keber points out that there are two watermarks in the manuscript. On folios 3 and 4 there is a teardrop bisected by a cross and roman numeral IV; see Keber, *Codex Telleriano-Remensis,* 123.

7. Studying watermarks has become important for tracking movements of political forces, trade, and influence in the early modern world, particularly between Latin America and Europe; see Fryer, "Spanish and Italian Watermarks"; Arnold, "Paper Ties to Land"; Wroth, *Some Reflections.*

8. In their work on legal documents in New Spain, Joanne Rappaport and Tom Cummins maintain that legal documents were primary instruments in "transforming native perceptions of time, space, and the discourses of power"; Rappaport and Cummins, *Beyond the Lettered City,* 4.

9. On the grid in a colonial context, see Baird, "Reordering of Space." On material encounters, see de Looze, "Transatlantic Textuality."

10. Bartolo da Sassoferrato, *Grammar of Signs,* 149. Ibid., 113.

11. Barrett, "Early European Papermaking Methods," 7.

12. Briquet, *Filigranes,* 1:10.

13. Ibid.

14. Walsh, "Medieval Merchant's Mark," 3.

15. Richards, *Florentine Merchants.*

16. Stevenson, "Briquet and the Future of Paper Studies," xlv.

17. Podgorny, "Toward a Bureaucratic History," 52; Siegert, *Passagiere und Papiere.* In 1636 the Spanish crown instituted *papel sellado,* with a governmental seal required for all administrative documents in its territories.

18. Perhaps the crown wanted to limit the making of paper in New Spain because it was a lucrative export; there were similar prohibitions against growing olives and grapes in the Americas because the crown wanted to keep the revenues generated from the taxes. See Stohr, "Watermarks in Venezuelan Documents."

19. The literature on land rights in New Spain is vast. For an overview, see Simpson, *Exploitation of Land;* Pagden, "Dispossessing the Barbarian"; Gibson, "Land"; Capdequi, *España en América.* On the history of land grants and paper, see Mundy, *Mapping of New Spain;* Russo, *Untranslatable Image;* Leibsohn and Pillsbury, "Writing the Land." As Lauren Benton points out, "The establishment of European courts of law and jurisdiction in New Spain raised the possibility of shared identity as the colonizers and the colonized occupied the status of subjects before the law, and it opened the way for the colonizers and the colonized to act in similar functions within the law—as litigants, advocates, witnesses, judges." Benton, *Search for Sovereignty,* 12.

20. Kellogg, *Law and Transformation,* xxii.

21. Valle, *Memorial;* Brotherston, *Painted Books from Mexico,* 154–76.

22. Valle, *Memorial,* 69–71.

23. Locke, *Essay Concerning Human Understanding,* 2:1, 104. Locke's epistemology of the mind as a tabula rasa already has received extensive criticism. Perhaps the most important critique, leveled by Richard Rorty, argued that Locke isolated the individual from shared experience and denied the effect of social practices and language in forming knowledge; see Rorty, *Philosophy and the Mirror of Nature.* For a countering of Rorty's reading, in which Locke "engages contingency, affective attachment, and imaginative language to critique existing authority and reconceive the subject," see Shanks, *Authority Figures,* 23.

24. Plato, *Plato's Theaetetus,* trans. David Bostock (Oxford: Clarendon, 1988),191c–196b. Aristotle, *De Anima,* trans. R. D. Hicks (Amsterdam: Adolf M. Hakkert, 1965), 430a, 424a.

25. Plato, *Plato's Theaetetus,* 191c–196b, 177.

26. Aristotle, *De Anima,* 430a, 424a, 105.

27. On the history of wax tablets, see Rouse and Rouse, "Vocabulary of Wax Tablets," 220–30.

28. Francis Bacon, "The Plan of the Work," in *Collected Works of Francis Bacon,* 4:27.

29. Hobbes, *Leviathan,* 289. For an overview of Locke's tabula rasa in the context of late seventeenth-century empiricism, see Wood, "Tabula Rasa."

30. Bacon, "Plan of the Work," 27.

31. Bacon, *New Organon,* 151–52.

32. On Locke and God-as-maker, see Tully, *Discourse on Property,* 36.

33. Ibid., 37.

34. Although Locke attempted to separate English colonists from the Spanish conquistadores, he employed many of the same legal arguments as Spanish jurists in regards to defending their right to this territory; see Herzog, "Did European Law Turn American?" Locke also was influenced by many of the narratives that he read from New Spain; see Castro-Klarén and Fernández, "Locke and Inca Garcilaso"; Arneil, *John Locke and America.* Locke, *Second Treatise of Government,* 5:42, 26.

35. Locke, *Second Treatise of Government,* 5:37, 24.

36. Ibid., 5:32, 21.

37. Ibid., 5:29, 29.

38. Siegert, *Cultural Techniques,* 12. See also Karla Mallette, "Boustrophedon: Towards a Literary Theory of the Mediterranean," in *A Sea of Languages: Rethinking the Arabic Role in Medieval Literary History* (Toronto: University of Toronto Press, 2013), 254–66.

39. Locke, *Essay Concerning Human Understanding,* 2:10:4, 151.

Conclusion

1. Silver, *Peasant Scenes and Landscapes,* 201–2.

2. Biörklund published an inventory of a VOC ship cargo from 1643 and 1644 that records carrying Japanese paper; see George Biörklund, "Old Paper," in *Rembrandt's Etchings, True and False: A Summary Catalogue* (New York: Museum Books, 1968), 172.

3. The exhibition was *Rembrandt's Etchings and Japanese Echizen Paper,* Rembrandt House Museum, Amsterdam, June 12–September 20, 2015. The Japanese origins for all of Rembrandt's Asian papers remain inconclusive. As a technical study at the National Gallery of Victoria revealed, some of Rembrandt's papers are made from bamboo and could also be from China; see Jacobus van Breda, "Rembrandt Etchings on Oriental Papers: Papers in the Collection of the National Gallery of Victoria," *Art Bulletin of Victoria* 38 (1997): 25–38.

4. Akira Kofuku, "Rembrandt's Prints on Asian Paper and Their Reception," in *Rembrandt: The Quest for Chiaroscuro* (Tokyo: National Museum of Western Art and Nippon Television Network, 2011), 129–38.

5. The ship of 1644 carried Japanese paper not only to the Netherlands but also to India. See Biörklund, "Old Paper," 172.

6. Leonard J. Slatkes, *Rembrandt and Persia* (New

York: Abaris, 1983), 13–58; Nicola Courtright, "Origins and Meaning of Rembrandt's Late Drawing Style," *Art Bulletin* 78, no. 3 (1996): 485–510.

7. Slatkes, *Rembrandt and Persia,* 25.

8. Martin Royalton-Kisch, *Drawings by Rembrandt and His Circle in the British Museum* (London: British Museum, 1992), 141–44.

9. Schrader, ed., *Rembrandt and the Inspiration of India.*

10. Guy, *Palm-Leaf and Paper,* 12.

11. Chandra, *Jain Miniature Paintings,* 70–74.

12. The literature on the hybridity of Mughal illumination is extensive; see Aitken, *Intelligence of Tradition;* Juneja, "Circulations and Beyond"; Singh, *Mughal Birds.*

13. On Rembrandt and Eastern curiosities, see Stephanie Schrader, "Rembrandt and the Mughal Line: Artistic Inspiration in the Global City of Amsterdam," in Schrader, *Rembrandt and the Inspiration of India,* 5–27.

14. Adrian Eeles, "Rembrandt's 'Ecce Homo': A Census of Impressions," *Print Quarterly* 15, no. 3 (1998): 290–96; Christopher White, *Rembrandt as an Etcher: A Study of the Artist at Work* (New Haven: Yale University Press, 1999).

15. Robert Fucci, *Rembrandt's Changing Impressions* (Cologne: Buchhandlung Walther König, 2015).

16. Anaïs Nin, "Ragtime," in *Anaïs Nin Reader,* ed. Philip K. Jason (Chicago: Swallow, 1973), 107.

17. Ibid., 108.

18. Ibid., 109.

19. On waste, recycling, reuse, and print, see also Leah Price, "The Book as Waste," in *How to Do Things with Books in Victorian Britain* (Princeton: Princeton University Press, 2012), 219–57. On the dynamics of recycling and use of printed texts, see William H. Sherman, *Used Books: Marking Readers in Renaissance England* (Philadelphia: University of Pennsylvania Press, 2008). As Sherman notes: "Printed images and texts were part of a dynamic ecology of use and reuse, leading to transformation and destruction as well as to preservation" (6).

BIBLIOGRAPHY

Abbott, Nabia. "A Ninth-Century Fragment of the 'Thousand Nights': New Light on the Early History of the Arabian Nights." *Journal of Near Eastern Studies* 8, no. 3 (1949): 129–64.

Abulafia, David. "Islam in the History of Early Europe." *European Review* 5 (1997): 241–56.

———. "The Role of Trade in Muslim–Christian Contact during the Middle Ages." In *Mediterranean Encounters: Economic, Religious, Political, 1100–1550*, 1–24. Aldershot, Eng.: Ashgate, 2000.

Abu-Lughod, Janet L. *Before European Hegemony: The World System, A.D. 1250–1350*. New York: Oxford University Press, 1989.

Ainsworth, Maryan W. "Review: *Early Netherlandish Drawings from Jan van Eyck to Hieronymus Bosch* by Fritz Koreny, Erwin Pokorny, and Georg Zeman [exhibition catalogue]." *Master Drawings* 41, no. 3 (2003): 305.

Ainsworth, Peter F. *Jean Froissart and the Fabric of History: Truth, Myth, and Fiction in the "Chroniques."* Oxford: Clarendon, 1990.

———. "Patterns of Transmission in the *Chroniques*." In *Froissart across the Genres*, 15–39. Gainesville: University Press of Florida, 1998.

Aitken, Molly Emma. *The Intelligence of Tradition in Rajput Court Painting*. New Haven: Yale University Press, 2010.

Alain of Lille. *The Plaint of Nature*. Trans. James J. Sheridan. Toronto: Pontifical Institute of Medieval Studies, 1980.

Alexander, Jonathan J. G. "Facsimiles, Copies, and Variations: The Relationship to the Model in Medieval and Renaissance Illuminated Manuscripts." *Studies in the History of Art: Retaining the Original: Multiple Originals, Copies, and Reproductions* 20 (1989): 61–72.

Amari, Michele. *I diplomi arabi del R. Archivio Fiorentino*. Florence: Tipi di F. Le Monnier, 1863–67.

Ames-Lewis, Francis. "Drawing for Panel-Painting in Trecento Italy: Reflections on Workshop Practice." *Apollo* (June 1992): 353–61.

———. *Drawing in Early Renaissance Italy*. New Haven: Yale University Press, 1981.

Amith, Jonathan D., ed. *La Tradición del amate: Innovación y protesta en el arte mexicano* (*The Amate Tradition: Innovation and Dissent in Mexican Art*). Mexico City: La Casa de las Imágenes; Chicago: Mexican Fine Arts Center Museum, 1995.

Antenhofer, Christina. "Die Gonzaga und Mantua. Kommunikation als Mittel der fürstlichen Herrschaft in der Stadt." In *Kommunikation in mittelalterlichen Städten*, ed. Jörg Oberste, 29–50. Regensburg, Ger.: Schnell und Steiner, 2007.

Anzelewsky, Fedja. *Albrecht Dürer: Das malerische Werk*. Berlin: Deutscher Verlag für Kunstwissenschaft, 1971.

Aristotle. *De Anima*. Trans. R. D. Hicks. Amsterdam: Adolf M. Hakkert, 1965.

Arneil, Barbara. *John Locke and America: The Defense of Colonialism*. Oxford: Clarendon, 1996.

Arnold, Philip. "Paper Rituals and the Mexican Landscape." In *Representing Ritual: Performance, Text, and Image in the Work of Sahagún*, ed. David Carrasco and Eloise Quiñones Keber, 227–50. Boulder: University Press of Colorado, 2002.

———. "Paper Ties to Land: Indigenous and Colonial Material Orientations in the Valley of Mexico." *History of Religions* 35, no. 1 (1995): 27–60.

Ashcroft, Jeffrey. *Albrecht Dürer: Documentary Biography*. New Haven: Yale University Press, 2017.

Assemanus, Giuseppe Simone. *Bibliotheca orientalis Clementino-vaticana.* Rome: Sacrae Congregationis de Propaganda Fide, 1719–28.

Bacon, Francis. *The Collected Works of Francis Bacon.* Ed. James Spedding, Robert Leslie Ellis, and Douglas Denon Heath. London: Longman, 1858–74.

———. *The New Organon.* Ed. Lisa Jardine and Michael Silverthorne. Cambridge: Cambridge University Press, 2000.

Bagnoli, Alessandro. "I tempi della Maestà. Il restauro e le nuove evidenze." In *Simone Martini, atti del convegno Siena, a cura di. Luciano Bellosi, 27–29 Marzo 1985,* 109–18. Florence: Centro Di, 1988.

Baird, Ellen T. "The Reordering of Space in Sixteenth-Century Mexico: Some Implications of the Grid." In *Painted Books and Indigenous Knowledge in Mesoamerica,* ed. Elizabeth Hill Boone, 289–300. New Orleans: Middle American Research Institute, 2005.

Baker, Don. "Arab Papermaking." *Paper Conservator* 15, no. 1 (1991): 28–35.

———. "A Note on the Expression: A Manuscript on Oriental Paper." *Manuscripts of the Middle East* 4 (1989): 67–68.

Bambach, Carmen C. *Drawing and Painting in the Italian Renaissance Workshop: Theory and Practice, 1300–1600.* New York: Cambridge University Press, 1999.

———. "The Purchases of Cartoon Paper for Leonardo's 'Battle of Anghiari' and Michelangelo's 'Battle of Cascina.'" *I Tatti Studies: Essays in the Renaissance* 8 (1999): 105–33.

Barbier, Frédéric. *Gutenberg's Europe: The Book and the Invention of Western Modernity.* Trans. Jean Birrell. Cambridge: Polity, 2017.

Barkan, Leonard. *Michelangelo: A Life on Paper.* Princeton: Princeton University Press, 2011.

Barrett, Timothy. "Early European Papermaking Methods, 1400–1800." *Paper Conservator* 13, no. 1 (2010): 7–27.

———. *Papermaking: Traditions, Tools, and Techniques.* Seattle: Legacy, 2018.

———. "The Woman Who Invented Notepaper: Towards a Comparative Historiography of Paper and Print." *Journal of the Royal Asiatic Society* 21, no. 2 (2011): 209–10.

Barthes, Roland. "The Plates of the Encyclopedia." In *A Barthes Reader,* ed. Susan Sontag, 218–35. New York: Hill and Wang, 1982.

Bartolo da Sassoferrato. *A Grammar of Signs: Bartolo da Sassoferrato's "Tract on Insignia and Coats of Arms."* Trans. Osvaldo Cavallar, Susanne Degenring, and Julius Kirshner. Berkeley: Robbins Collection, University of California at Berkeley, 1995.

Bat-Yehouda, Monique Zerdoun, ed. *Le papier au Moyen Âge: Histoire et techniques.* Turnhout, Belg.: Brepols, 1999.

Batalla Rosado, Juan José. "The Scribes Who Painted the Matrícula de Tributos and the Codex Mendoza." *Ancient Mesoamerica* 18 (2007): 31–51.

Beit-Arié, Malachi. "Oriental Arabic Paper." *Gazette du livre medieval* 28, no. 1 (1996): 9–12.

Benton, Lauren. *A Search for Sovereignty: Law and Geography in European Empires, 1400–1900.* Cambridge: Cambridge University Press, 2010.

Berdan, Frances F. "The Imperial Tribute Roll of the Codex Mendoza." In *The Codex Mendoza,* ed. Frances F. Berdan and Patricia R. Anawalt, 1:55–80. Berkeley: University of California Press, 1992.

Berdan, Frances F., and Jacqueline de Durand-Forest. *Matrícula de Tributos (Códice de Moctezuma): Museo Nacional de Antropologiá, México (Cód. 35–52).* Graz, Austria: Akademsche Druck- u. Verlagsanstalt, 1980.

Bernard, Frédéric, and Bernard Picart. *Cérémonies et coutumes religieuses, de tous les peuples du monde.* Amsterdam: Chez J. F. Bernard, 1723–43.

Bidwell, John. "The Study of Paper as Evidence, Artefact, and Commodity." In *The Book Encompassed: Studies in Twentieth-Century Bibliography,* ed. Peter Davison, 69–82. Cambridge: Cambridge University Press, 1998.

Blair, Ann. "Note-Taking as an Art of Transmission." *Critical Inquiry* 31 (2004): 85–107.

Blair, Sheila. "Color and Gold: The Decorated Papers Used in Manuscripts in Later Islamic Times." *Muqarnas* 17 (2000): 24–36.

Bleichmar, Daniela. "History in Pictures: Translating the *Codex Mendoza.*" *Art History* 38, no. 4 (2015): 682–701.

Bloch, Marc. "The Advent and Triumph of the Watermill." In *Land and Work in Medieval Europe: Selected Papers by Marc Bloch*, 136–68. London: Routledge, 1967.

———. "Avènement et conquêtes du Moulin à eau." *Annales d'histoire économique et sociale* 7 (1935): 538–63.

Bloom, Jonathan. *Paper before Print: The History and Impact of Paper in the Islamic World.* New Haven: Yale University Press, 2001.

Blum, André. *On the Origin of Paper.* New York: Bowker, 1934.

Bohrer, Frederick N. "Photographic Perspectives: Photography and the Institutional Formation of Art History." In *Art History and Its Institutions: Foundations of a Discipline*, ed. Elizabeth Mansfield, 246–59. London: Routledge, 2002.

Boon, K. G. "Over grisaille en zilverstift. Stromingen in de Franse tekenkunst rond 1400." *Maandblad voor Beeldende Kunsten* 26 (1950): 263–73.

Boone, Elizabeth Hill. *Stories in Red and Black: Pictorial Histories of the Aztecs and Mixtecs.* Austin: University of Texas Press, 2000.

Boone, Elizabeth Hill, and Walter D. Mignolo, eds. *Writing without Words: Alternative Literacies in Mesoamerica and the Andes.* Durham, N.C.: Duke University Press, 1994.

Borsook, Eve. *The Mural Painters of Tuscany from Cimabue to Andrea de Sarto.* London: Phaidon, 1960.

Bourlet, Caroline. "Les tabletiers parisiens à la fin du moyen âge." In *Les tablettes à écrire de l'antiquité à l'époque modern*, ed. Élisabeth Lalou, 338–41. Turnhout, Belg.: Brepols, 1992.

Bower, Peter. "The White Art: The Importance of Interpretation in the Analysis of Paper." In *Looking at Paper: Evidence and Interpretation. Symposium Proceedings, Toronto 1999*, ed. John Slavin, 5–16. Ottawa: Canadian Conservation Institute, 2001.

Bratu, Christian. "'Je, acteur de ce livre': Authorial Persona and Authority in French Medieval Histories and Chronicles." In *Authorities in the Middle Ages: Influence, Legitimacy, and Power in Medieval Society*, ed. Sini Kangas, Mia Korpiola, and Tuija Ainonen, 183–204. Berlin: De Gruyter, 2013.

Briquet, Charles-Moïse. *Briquet's Opuscula: The Complete Works of Dr. C. M. Briquet without Les filigranes.* Hilversum, Neth.: Paper Publications Society, 1955.

———. *Les filigranes: Dictionnaire historique des marques du papier dès leur apparition vers 1282 jusqu'en 1600.* Leipzig, Neth.: Hiersemann, 1923.

———. *Papiers et filigranes des archives des Gênes 1154 à 1700.* Geneva: H. Georg, 1888.

———. "Recherches sur les premiers papiers employés en occident et en orient du Xe au XIVe siècle." In *Mémoires de la societé nationale des antiquaries de France*, 133–205. Paris: La Société, 1871–.

Brotherston, Gordon. *Painted Books from Mexico: Codices in UK Collections and the World They Represent.* London: Published by the Trustees of the British Museum, 1995.

Brownlee, Kevin. *Poetic Identity in Guillaume de Machaut.* Madison: University of Wisconsin Press, 1984.

Brückle, Irene. "Blue-Colored Paper in Drawings." *Drawing* 15, no. 4 (1993): 73–77.

———. "The Historical Manufacture of Blue-Coloured Paper." *Paper Conservator* 17, no. 1 (1993): 20–31.

Buchthal, Hugo. *The Musterbuch of Wolfenbüttel and Its Position in the Art of the Thirteenth Century.* Vienna: Österreichischen Akademie der Wissenschaften, 1979.

Buck, Stephanie. "Tradition als Herausforderung: Dürers früheste Figurenstudie." In *Der frühe Dürer*, ed. Daniel Hess and Thomas Eser, 90–100. Nuremberg: Germanischen Nationalmuseum, 2012.

Burns, Robert I. "Paper Comes to the West, 800–1400." In *Europäische Technik im Mittelalter, 800–1400*, ed. U. Lindgreen, 413–22. Berlin: Mann, 1996.

———. "The Paper Revolution in Europe: Crusader Valencia's Paper Industry: A Technological and Behavioral Breakthrough." *Pacific Historical Review* 50, no. 1 (1981): 1–30.

———. *Society and Documentation in Crusader Valencia.* Princeton: Princeton University Press, 1985.

Cabañas, Antoni Aragó, and José Trenchs. "Los registros de cancilleria de la coronoa de Aragón (Jaime I y Pedro II) y los registors pontificos." *Annali della scuola speciale per archivisti e bibliotecari dell'università di Roma (III Congresso internazionale di diplomatica)* 12 (1972): 26–39.

Calhoun, Joshua. "The Word Made Flax: Cheap Bibles, Textual Corruption, and the Poetics of Paper." *PMLA* 126, no. 2 (2011): 327–44.

Calvo, Hortensia. "The Politics of Print: The History of the Book in Early Spanish America." *Book History* 6 (2003): 277–305.

Camille, Michael. *The Medieval Art of Love: Objects and Subjects of Desire.* New York: Abrams, 1998.

Cammarosano, Paolo. *Italia medievale: Struttura e geografia delle fonti scritte.* Rome: Nuova Italia scientifica, 1991.

Campbell, Angela. "Finding Folds: Albrecht Dürer's Meisterstiche Papers." *Print Quarterly* 29, no. 4 (2012): 405–11.

Capdequi, J. M. Ots. *España en América: El régimen de tierras en la época colonial.* Mexico City: Fondo de Cultura Económica, 1959.

Carter, Nicholas P., and Jeffrey Dobereiner. "Multispectral Imaging of an Early Classic Maya Codex Fragment from Uaxactun, Gautemala." *Antiquity* 90 (2016): 711–25.

Casiri, Miguel. *Bibliotheca Arabico-Hispana Escurialensis.* Madrid: Antonius Perez de Soto, 1760–70.

Castro-Klarén, Sara, and Christian Fernández. "Locke and Inca Garcilaso." In *Inca Garcilaso and Contemporary World-Making,* 269–96. Pittsburgh: University of Pittsburgh Press, 2015.

Cennini, Cennino. *Cennino Cennini's "Il libro dell'arte": A New English Translation and Commentary with Italian Transcription.* Ed. Lara Broecke. London: Archetype, 2015.

Cerquiglini-Toulet, Jacqueline. *The Color of Melancholy: The Uses of Books in the Fourteenth Century.* Trans. Lydia G. Cochrane. Baltimore: Johns Hopkins University Press, 1997.

Chambers, D. S. "A Condottiere and His Books: Gianfrancesco Gonzaga (1446–96)." *Journal of the Warburg and Courtauld Institutes* 70 (2007): 33–97.

Chandra, Moti. *Jain Miniature Paintings from Western India.* Ahmedabad: S. W. Nawab, 1949.

Chapman, Hugo. *Fra Angelico to Leonardo: Italian Renaissance Drawings.* London: British Museum, 2010.

Châtelet, Albert. "Un artiste à la cour de Charles VI. A propos d'un carnet d'esquisses de XIVe siècle conserve à la Pierpont Morgan Library." *L'oeil* 216 (1972): 116–21.

Clemens, Robert J. "Michelangelo on Effort and Rapidity in Art." *Journal of the Warburg and Courtauld Institutes* 17 (1954): 301–10.

Clendinnen, Inga. *Ambivalent Conquests: Maya and Spaniard in Yucatan, 1517–1570.* Cambridge: Cambridge University Press, 1987.

Cole, Arthur Harrison. *The Handicrafts of France as Recorded in the "Description des arts et des métiers" 1761–1788.* Boston: Baker Library, 1952.

Continelli, Luisa. *L'Archivio dell'Ufficio dei Memoriali: Inventario.* Vol. 1, *Memoriali, 1265–1436.* Bologna: Bologna University Press, 2008.

Conzatti, Emilia Seemann. *Usos del papel en el Calendario Ritual Mexicana.* Mexico City: Instituto Nacional de Antropología e Historia, 1990.

Corsi, Elisabetta. "Scholars and Paper-Makers: Paper and Paper-Manufacture According to Tu Long's Notes on Paper." *Rivista degli studi orientali* 65 (1991): 69–107.

Cossar, Roisin. *Clerical Households in Late Medieval Italy.* Cambridge, Mass.: Harvard University Press, 2017.

D'Anghiera, Peter Martyr. *De Orbe Novo.* Trans. Francis Augustus MacNutt. New York: Knickerbocker, 1912.

Dannenfeldt, Karl H. "The Renaissance Humanists and the Knowledge of Arabic." *Studies in the Renaissance* 2 (1955): 96–117.

Daybell, James. *The Material Letter in Early Modern England: Manuscript Letters and the Culture of Letter-Writing, 1512–1635.* Basingstoke, Eng.: Palgrave Macmillan, 2012.

Degenhart, Bernhard, and Annegrit Schmitt. *Corpus der italienischen Zeichnungen, 1300–1450.* Berlin: Mann, 1968.

Dekeyzer, Brigitte. "Word and Image: Foundations of the Medieval Manuscript." In *Medieval Mastery: Book Illumination from Charlemagne to Charles the Bold, 800–1475,* 25–34. Turnhout, Belg.: Brepols, 2002.

Déroche, François. *Islamic Codicology: An Introduction to the Study of Manuscripts in Arabic Script.* London: Al-Furqan Islamic Heritage

Foundation, 2005.

Derrida, Jacques. *Archive Fever: A Freudian Impression.* Trans. Eric Prenowitz. Chicago: University of Chicago Press, 1996.

———. *Paper Machine.* Trans. Rachel Bowlby. Stanford: Stanford University Press, 2005.

Douglas, Eduardo de J. *In the Palace of Nezahualcoyotl: Painting Manuscripts, Writing the Pre-Hispanic Past in Early Colonial Period Tetzcoco, Mexico.* Austin: University of Texas Press, 2010.

Dumas, Maurice, and René Tresse. "La description des arts et métiers de l'Académie des Sciences et le sort de ses planches gravées en taille douce." *Revue d'histoire des science et de leurs applications* (1954): 163–71.

Dürer, Albrecht. *Schriftlicher Nachlass.* Ed. Hans Rupprich. Berlin: Deutscher Verein für Kunstwissenschaft, 1956.

Earp, Lawrence. "Machaut's Role in the Production of Manuscripts of His Works." *American Musicological Society* 42, no. 3 (1989): 461–503.

Echevarría, Roberto González. *Myth and Archive: A Theory of Latin American Narrative.* Durham, N.C.: Duke University Press, 1998.

Ellis, Margaret Holben, ed. *Historical Perspectives in the Conservation of Works of Art on Paper.* Los Angeles: Getty Conservation Institute, 2014.

Emison, Patricia. "Grazia." *Renaissance Studies* 5, no. 4 (1991): 427–60.

Estève, Jean-Louis. "Le zigzag dans les papiers arabes. Essai d'explication." *Gazette du livre médiéval* 38 (2001): 40–49.

Fairbanks, Theresa. *Papermaking and the Art of Watercolour in Eighteenth-Century Britain: Paul Sandby and the Whatman Paper Mill.* New Haven: Yale University Press, 2006.

Febvre, Lucien, and Henri-Jean Martin. *The Coming of the Book: The Impact of Printing, 1450–1800.* Trans. David Gerard. London: Humanities, 1976.

Ferrari, Daniela. "The Gonzaga Archives of Mantua and Their Rearrangements over the Centuries, along with an Overview of Archival Materials on Mantuan Jewry." In *Rabbi Judah Moscato and the Jewish Intellectual World of Mantua in 16th–17th Century,* ed. Giuseppe Veltri and Gianfranco Miletto, 159–74. Leiden, Neth.: Brill, 2012.

Filelfo, Francesco. *On Exile.* Ed. Jeroen de Keyser. Trans. W. Scott Blanchard. Cambridge, Mass.: Harvard University Press, 2013.

Foucault, Michel. *Aesthetics, Method, and Epistemology.* Ed. James D. Faubion. Trans. Robert Hurley. New York: New Press, 1998.

———. *The Archaeology of Knowledge.* Trans. A. M. Sheridan Smith. New York: Pantheon, 1972.

———. *The Order of Things: An Archaeology of the Human Sciences.* New York: Vintage, 1970.

Friedrich, Markus. "Archival Practices: Producing Knowledge in Early Modern Repositories of Writing." In *Praktiken der frühen Neuzeit: Akteure, Handlungen, Artefakte,* ed. Arndt Brendecke, 468–72. Cologne: Böhlau, 2015.

Friend, Albert Matthias. "The Portraits of the Evangelists in Greek and Latin Manuscripts, Part I." *Art Studies* 5 (1927): 115–50.

———. "The Portraits of the Evangelists in Greek and Latin Manuscripts, Part II." *Art Studies* 7 (1929): 3–46.

Frinta, Mojimír S. "Reviewed Work(s): Das Skizzenbuch des 'Jaques Daliwe' by Helga Kreuter-Eggemann." *Art Bulletin* 52, no. 1 (1970): 100–102.

Froissart, Jean. *An Anthology of Narrative and Lyric Poetry.* Ed. and trans. Kristen M. Figg and R. Barton Palmer. New York: Routledge, 2001.

———. *Chroniques.* Ed. Andrée Duby. Paris: Stock, 1997.

———. *Chroniques.* Ed. Peter Ainsworth and George T. Diller. Paris: Librairie générale française, 2001–ca. 2004.

———. *Chroniques: Début du premier livre/éd. du manuscript de Rome Reg. lat. 869.* Ed. George T. Diller. Geneva: Droz, 1972.

———. *La prison amoureuse–The Prison of Love.* Ed. and trans. Laurence De Looze. New York: Garland, 1994.

Fry, Roger. "On a Fourteenth-Century Sketchbook." *Burlington* 10 (1906): 31–38.

Fryer, Celia A. "Spanish and Italian Watermarks in Colonial Guatemalan Books." In *Puzzles in Paper: Concepts in Historical Watermarks: Essays from the International Conference on Watermarks in Roanoke, Virginia,* ed. Daniel Moser, Michael Saffle, and Ernest W. Sullivan II, 37–55. New Castle, Del.: Oak Knoll, 2000.

Gacek, Adam. "On the Making of Local Paper: A

Thirteenth-Century Yemeni Recipe." *La tradition manuscrite en écriture arabe* 99–100 (2002): 79–93.
Gasparinetti, A. F. "Ein altes Statut von Bologna über die Herstellung und den Handel von Papier." *Papiergeschichte* 6, no. 3 (1956): 45–47.
Geymonat, Ludovico V. "Drawing, Memory and Imagination in the Wolfenbüttel Musterbuch." In *Mechanisms of Exchange: Transmission in Medieval Art and Architecture of the Mediterranean ca. 1000–1500*, ed. Heather E. Grossman and Alicia Walker, 220–85. Leiden, Neth.: Brill, 2012.
Ghobrial, John-Paul. "The Archive of Orientalism and Its Keepers: Reimagining the Histories of Arabic Manuscripts in Early Modern Europe." *Past and Present: Supplement* 11 (2016): 90–111.
Gibson, Charles. "Land." In *The Aztecs under Spanish Rule: A History of the Indians of the Valley of Mexico 1519–1810*, 257–99. Stanford: Stanford University Press, 1964.
Gitelman, Lisa. *Paper Knowledge: Toward a Media History of Documents*. Durham, N.C.: Duke University Press, 2014.
Goitein, Shelomo Dov. *A Mediterranean Society: The Jewish Communities of the Arab World as Portrayed in the Documents of the Cairo Geniza*. 6 vols. Berkeley: University of California Press, 1967–93.
Golas, Peter J. "'Like Obtaining a Great Treasure': The Illustrations in Song Yingxing's *The Exploitation of the Works of Nature*." In *Graphics and Text in the Production of Technical Knowledge in China*, ed. Francesca Bray, Vera Dorofeeva-Lichtmann, and Georges Métailié, 569–614. Leiden, Neth.: Brill, 2007.
Grabar, Oleg. *The Illustrations of the Maqamat*. Chicago: University of Chicago Press, 1984.
Grazia, Margreta de, and Peter Stallybrass. "The Materiality of the Shakespearean Text." *Shakespeare Quarterly* 44, no. 3 (1993): 255–83.
Grebe, Anja. "Albrecht Dürers 'Kunstbücher': Ordnungssysteme frühneuzeitlicher Grafiksammlungen und die Anfänge des Catalogue raisonné." *Marburger Jahrbuch für Kunstwissenschaft* 39 (2012): 27–75.
Gruber, Bettina, and Jürgen Müller. "Künstler und Gesellschaft in Renaissance und Manierismus." In *Handbuch Rhetorik der Bildenden Künste*, ed. Wolfgang Brassat, 229–52. Berlin: Walter de Gruyter, 2017.
Guichard, Pierre. "Du parchemin au papier." In *Comprendre le XIIIe siècle: études offertes à Marie-Thérèse Lorcin*, ed. Pierre Guichard and Danièle Alexandre-Bidon, 185–200. Lyon: Presses universitaires de Lyon, 1995.
Guy, John. *Palm-Leaf and Paper: Illustrated Manuscripts of India and Southeast Asia*. Melbourne: National Gallery of Victoria, 2012.
Hagen, Victor Wolfgang von. *The Aztec and Maya Papermakers*. New York: J. J. Augustin, 1943.
Halde, Jean-Baptiste du. *Description de la Chine*. The Hague: Henri Scheurleer, 1736.
———. *A Description of the Empire of China and Chinese-Tartary: Together with the Kingdoms of Korea, and Tibet: Containing the Geography and History (Natural as Well as Civil) of Those Countries*. London: Printed by T. Gardner, 1738–41.
Halevi, Leo. "Christian Impurity versus Economic Necessity: A Fifteenth-Century Fatwa on European Paper." *Speculum* 83, no. 4 (2008): 917–45.
Herzog, Tamar. "Did European Law Turn American? Territory, Property, and Rights in an Atlantic World." In *New Horizons in Spanish Colonial Law: Contributions to Early Modern Legal History*, ed. Thomas Duve and Heikki Pihlajamäki, 75–95. Frankfurt: Max Planck Institute for European Legal History, 2015.
Hills, Richard, L. "Early Spanish Papers." *IPH Information* 4 (1989): 165–69.
———. "Papermaking Stampers: A Study in Technological Diffusion." *International Association of Paper Historians Yearbook* 5 (1984): 67–76.
———. "A Technical Revolution in Papermaking, 1250–1350." In *Looking at Paper: Evidence and Interpretation: Symposium Proceedings, Toronto 1999*, ed. John Slavin, 105–11. Ottawa: Canadian Conservation Institute, 2001.
Hobbes, Thomas. *Leviathan*. Ed. A. P. Martinich and Brian Battiste. Peterborough, Ont.: Broadview, 2011.
Holcomb, Melanie. *Pen and Parchment: Drawing in the Middle Ages*. New Haven: Yale University Press, 2009.

Howard, Deborah. *Venice and the East: The Impact of the Islamic World on Venetian Architecture.* New Haven: Yale University Press, 2000.

Huard, Georges. "Les planches de l'Encyclopédie et celles de la Description des Arts et Métiers de l'Académie des Sciences." *Revue d'histoire des sciences et de leurs applications* (1951): 238–49.

Humbert, Geneviève. "Le manuscrit arabe et ses papiers." *Revue des mondes musulmans et de la Méditerrannée* 99–100 (2002): 55–77.

———. "Papiers non filigranés utilisés au proche-orient jusqu'en 1450: Essai de typologie." *Journal asiatique* 286, no. 1 (1998): 1–54.

Humfrey, Peter. "Dürer's *Feast of the Rose Garlands:* A Venetian Altarpiece." *Bulletin of the National Gallery in Prague* 1 (1991): 21–37.

Hunter, Dard. *Old Papermaking in China and Japan.* Chillicothe, Ohio: Mountain House, 1932.

———. *Papermaking: The History and Technique of an Ancient Craft.* New York: Knopf, 1943.

———. *Primitive Papermaking: An Account of a Mexican Sojourn and of a Voyage to the Pacific Islands in Search of Information, Implements, and Specimens Relating to the Making and Decorating of Bark-Paper.* Chillicothe, Ohio: Mountain House, 1927.

Huot, Sylvia. *From Song to Book: The Poetics of Writing in Old French Lyric and Lyrical Narrative Poetry.* Ithaca: Cornell University Press, 1987.

Innis, Harold. *A History of Communications: Paper and Printing.* Ed. William J. Buxton, Michael R. Cheney, and Paul Heyer. Lanham, Md.: Rowman and Littlefield, 2015.

Irigoin, Jean. "Les types des formes utilisés dans l'Orient méditerranéen (Syrie, Égypte) du XIe au XIVe siècle." *Papiergeschichte* 13 (1963): 18–21.

Israëls, Machtelt. *Sassetta's Madonna della Neve: An Image of Patronage.* Leiden, Neth.: Primavera, 2003.

Jaucourt, Louis de. "Papier (11:846)." In *Encyclopédie, ou dictionnaire raisonné des sciences, des arts et des métiers, etc.,* ed. Denis Diderot and Jean le Rond d'Alembert. University of Chicago: ARFTL Encyclopédie Project (Spring 2016 ed.), http://encyclopedie.uchicago.edu/.

Jenni, Ulrike. "Von mittelalterlichen Musterbuch zum Skizzenbuch der Neuzeit." In *Die Parler und der Schöne Stil 1350–1400: Europäische Kunst unter den Luxemburgern,* 3:139–41. Cologne: Museen der Stadt Köln, 1978–80.

———. "The Phenomena of Change in the Model-book Tradition around 1400." In *Drawings Defined,* ed. Walter Strauss and Tracie Felker, 35–47. New York: Abaris, 1987.

———. *Das Skizzenbuch des Jaques Daliwe: Kommentar zur Faksimileausgabe des Liber picturatus A 74 der Deutschen Staatsbibliothek Berlin/DDR.* Leipzig: VCH, 1987.

Jerome. *Biblia sacra: Iuxta vulgatam versionem.* Ed. Robertus Weber. Stuttgart: Württembergische Bibelanstalt, 1969.

———. *Letters and Select Works: A Select Library of Nicene and Post-Nicene Fathers of the Christian Church.* 2nd ser. Trans. Philip Schaff and Henry Wace. Grand Rapids, Mich.: Eerdmans, 1978–79.

Jones, Robert. "The Medici Oriental Press in Rome (1584–1614) and the Impact of Its Arabic Publications." In *The "Arabick" Interests of the Natural Philosophers in Seventeenth-Century England,* ed. G. A. Russell, 88–108. Leiden, Neth.: Brill, 1994.

———. "Piracy, War, and the Acquisition of Arabic Manuscripts in Renaissance Europe." *Manuscripts of the Middle East* 2 (1987): 96–110.

Juneja, Monica. "Circulations and Beyond—The Trajectories of Vision in Early Modern Eurasia." In *Circulations in the Global History of Art,* ed. Thomas DaCosta Kaufmann, Catherine Dossin, and Béatrice Joyeux-Prunel, 59–78. London: Ashgate, 2015.

Kaempfer, Engelbert. *The History of Japan.* London: Printed for the publisher, and sold by T. Woodward and C. Davis, 1727.

Kafka, Ben. *The Demon of Writing: Powers and Failures of Paperwork.* New York: Zone, 2012.

———. "Paperwork: The State of the Discipline." *Book History* 12 (2009): 340–53.

Karabacek, Joseph von. "Das Arabische Papier (eine historisch-antiquarische Untersuchung)." *Mittheilungen der Papyrussammlung Erzherzog Rainer* 2–3 (1887): 87–178.

Keber, Eloise Quiñones. *Codex Telleriano-Remensis: Ritual, Divination, and History in a Pictorial Aztec Manuscript.* Austin: University of Texas Press, 1995.

Kellogg, Susan. *Law and the Transformation of Aztec Culture, 1500–1700*. Norman: University of Oklahoma Press, 1995.

Kemperdick, Stephan. *Martin Schongauer.* Petersberg, Ger.: M. Imhof, 2004.

Kendrick, Laura. *The Game of Love: Troubadour Wordplay.* Berkeley: University of California Press, 1988.

Ketelaar, Eric. "Records Out and Archives In: Early Modern Cities as Creators of Records and as Communities of Archives." *Archival Science* 10 (2010): 201–210.

Köllner, Herbert. "Das Skizzenbuch des 'Jaques Daliwe' by Helga Kreuter-Eggemann." *Zeitschrift für Kunstgeschichte* 33 (1970): 68–75.

König, Eberhard. "How Did Illuminators Draw?" *Master Drawings* 41, no. 3 (2003): 216–27.

———. *Vom Psalter zum Stundenbuch: Zwei bedeutende Handschriften aus dem 14. Jahrhundert. Mit einem Versuch über das Phänomen Jacquemart de Hesdin.* Ramsen, Switz.: Antiquariat Bibermühle, Heribert Tenschert, 2015.

König, Eberhard, and Maria Theisen. *Wiener musterbuch.* Simbach am Inn, Ger.: Mueller und Schindler, 2012.

Koreny, Fritz. "Venice and Dürer." In *Renaissance Venice and the North: Crosscurrents in the Time of Bellini, Dürer, and Titian,* ed. Bernard Aikema and Beverly Louise Brown, 332–423. New York: Rizzoli, 2000.

Kotková, Olga, ed. *Albrecht Dürer: The Feast of the Rose Garlands, 1506–2006.* Prague: Nárdoni galerie v Praze, 2006.

Kreuter-Eggemann, Helga. *Das Skizzenbuch des "Jaques Daliwe."* Munich: Bruckmann, 1964.

Krill, John. *English Artist's Paper: Renaissance to Regency.* New Castle, Del.: Oak Knolls, 2001.

Laguna-Chevillotte, Agnieszka. "Le Christ-Verbe: remarques sur la fortune du theme dans la statuaire mariale de la mouvance royale." In *La creation artistique en France autour de 1400: actes du colloque international, École du Louvre, Musée des beaux-arts de Dijon, Université de Bourgogne,* 409–30. Paris: École du Louvre, 2006.

La Lande, Jérôme de. *Art de faire le papier: Descriptions des arts et des métiers, faites our approuvées par MM. de l'Académie royale des sciences.* Paris: Saillant et Nyon, 1761.

Landa, Diego de. *Landa's Relación de las cosas de Yucatán.* Trans. Alfred M. Tozzer. Cambridge, Mass.: Peabody Museum, 1941.

———. *Relación de las Cosas de Yucatán.* Newark: Juan de la Cuesta, 2011.

Landau, David, and Peter Parshall. *The Renaissance Print, 1470–1550.* New Haven: Yale University Press, 1994.

Leach, Bridget, and John Tait. "Papyrus." In *Ancient Egyptian Materials and Technology,* ed. Paul T. Nicholson and Ian Shaw, 227–53. Cambridge: Cambridge University Press, 2009.

Leach, Eva Elizabeth. *Guillaume de Machaut: Secretary, Poet, Musician.* Ithaca: Cornell University Press, 2011.

Leibsohn, Dana, and Joanne Pillsbury. "Writing the Land." *Script and Glyph: Pre-Hispanic History, Colonial Bookmaking and the "Historia Tolteca-Chichimeca": Studies in Pre-Columbian Art and Archaeology* 36 (2009): 62–92.

Lenz, Hans. *Historia del papel en México y cosas relacionadas, 1592–1950.* Mexico City: M. A. Porrúa, 1990.

———. *El papel indigena mexicano: Historia y supervivencia.* Mexico City: Editorial Cultura, 1950.

Levey, Martin. "Mediaeval Arabic Bookmaking and Its Relation to Early Chemistry and Pharmacology." *Transactions of the American Philosophical Society* 52, no. 4 (1962): 1–79.

Lightbrown, Ronald. *Mantegna: With a Complete Catalogue of the Paintings, Drawings and Prints.* Oxford: Phaidon; Oxfordshire: Christie's, 1986.

Lincoln, Evelyn. "Gospel Lessons: Arabic Printing at the Tipografia Medicea Oriental." *Art in Print* 4, no. 4 (2014): 4–10.

Lipshitz, Yair. "Performance as Profanation: Holy Tongue and Comic Stage in Tsahut bedihuta deqiddushin." *Renaissance Drama* 36/37 (2010): 127–57.

Locke, John. *An Essay Concerning Human Understanding.* Oxford: Clarendon, 1975.

———. *Second Treatise of Government.* Ed. C. B. Macpherson. Indianapolis: Hackett, 1980.

Looze, Laurence de. "Transatlantic Textuality: The Encounter between Spanish and Indigenous Textuality in the Pages of the Sixteenth-Century

'Codices Mestizos' of Mexico." *Mediterranean Studies* 14 (2005): 106–24.

Loveday, Helen. *Islamic Paper: A Study of the Ancient Craft.* London: Don Baker Memorial Fund, 2001.

Luber, Katherine Crawford. *Albrecht Dürer and the Venetian Renaissance.* Cambridge: Cambridge University Press, 2005.

Lucas, Adam Robert. "Industrial Milling in the Ancient and Medieval Worlds: A Survey of the Evidence for an Industrial Revolution in Medieval Europe." *Technology and Culture* 46 (2005): 1–30.

———. *Wind, Water, Work: Ancient and Medieval Milling Technology.* Leiden, Neth.: Brill, 2006.

Lugli, Emanuele. "Hidden in Plain Sight: The 'Pietre di Paragone' and the Preeminence of Medieval Measurements in Communal Italy." *Gesta* 49, no. 2 (2010): 77–95.

Lunning, Elizabeth. "Characteristics of Italian Paper in the Seventeenth Century." In *Italian Sketches of the Renaissance and Baroque,* ed. Sue Walsh and Richard W. Wallace, xxii–xliii. Boston: Museum of Fine Arts, 1989.

Machaut, Guillaume de. *"Dits" et "debats."* Geneva: Droz, 1979.

———. *Le livre dou voir dit* (*The Book of the True Poem*). Ed. Daniel Leech-Wilkinson. Trans. R. Barton Palmer. New York: Garland, 1998.

Mack, Rosamond. *Bazaar to Piazza: Islamic Trade and Italian Art, 1300–1600.* Berkeley: University of California Press, 2000.

Mack, Rosamond, and Mohamed Zakariya. 'The Pseudo-Arabic on Andrea del Verrocchio's David." *Artibus et historiae* 30, no. 60 (2009): 157–72.

Maffie, James. *Aztec Philosophy: Understanding a World in Motion.* Denver: University of Colorado Press, 2015.

Maginnis, Hayden B. J. "The Craftsman's Genius: Painters, Patrons and Drawings in Trecento Siena." In *The Craft of Art: Originality and Industry in the Italian Renaissance and Baroque Workshop,* ed. Andrew Ladis and Carolyn Wood, 25–47. Athens: University of Georgia Press, 1995.

———. *The World of the Early Sienese Painter.* University Park: Pennsylvania State University Press, 2001.

Maier, Michael. *Lusius serius, or Serious Passe-time: A Philosophical Discourse Concerning the Superiority of Creatures under Man.* London: Printed for Humphrey Moseley, 1654.

Maire-Vigueur, Jean-Claude. "Révolution documentaire et revolution scripturaire: Le cas de l'Italie medieval." *Bibliothèque de l'École des chartres* 153 (1995): 177–85.

Mak, Bonnie. *How the Page Matters.* Toronto: University of Toronto Press, 2011.

McGrady, Deborah. *Controlling Readers: Guillaume de Machaut and His Late Medieval Audience.* Toronto: University of Toronto Press, 2006.

———. "Machaut and His Material Legacy." In *A Companion to Guillaume de Machaut,* ed. Deborah McGrady and Jennifer Bain, 361–85. Leiden, Neth.: Brill, 2012.

McTigue, Clara de la Peña. "Paper and Papermaking in Spain." In *Renaissance to Goya: Prints and Drawing from Spain,* ed. Mark P. McDonald, 274–81. Burlington, Vt.: Lund Humphries, 2012.

Meder, Joseph. *The Mastery of Drawing.* Trans. Winslow Ames. New York: Abaris, 1978.

Meiss, Millard. *French Painting in the Time of Jean de Berry: The Late Fourteenth Century and the Patronage of the Duke.* London: Phaidon, 1967.

Milde, Wolfgang. "Zum Wolfenbütteler Musterbuch." In *Die Zeit der Staufer, V. Supplement: Vorträge und Forschungen,* ed. R. Hausherr and C. Väterlein, 331–33. Stuttgart: Württembergisches Landesmuseum, 1979.

Minnis, Alastair J. *Medieval Theory of Authorship: Scholastic Literary Attitudes in the Later Middle Ages.* London: Scholar, 1984.

Morrall, Andrew. "Dürer and Venice." In *The Essential Dürer,* ed. Larry Silver and Jeffrey Chipps Smith, 99–114. Philadelphia: University of Pennsylvania Press, 2010.

Muller, Norman. "Paper in Simone Martini's Frescoed *Maestà.*" *Quarterly* 82 (2012): 12–13.

Mundy, Barbara. *The Mapping of New Spain: Indigenous Cartography and the Maps of the Relaciones Geográficas.* Chicago: University of Chicago Press, 1996.

Nagel, Alexander. *Some Discoveries of 1492: Eastern Antiquities and Renaissance Europe.* Groningen, Neth.: Gerson Lectures Foundation, 2013.

———. "Twenty-Five Notes on Pseudoscript in Italian Art." *Res: Anthropology and Aesthetics* 59/60 (2011): 228–48.

Nelson, Robert S. "The Inspired or Inspiring Evangelist." In *The Iconography of the Preface and Miniature in the Byzantine Gospel Book*, 75–92. New York: New York University Press, 1980.

———. "The Slide Lecture, or the Work of Art 'History' in the Age of Mechanical Reproduction." *Critical Inquiry* 26, no. 3 (2000): 414–34.

Nichols, Stephen G. "Voice and Writing in Augustine and in the Troubadour Lyric." In *Vox Intexta: Orality and Textuality in the Middle Ages*, ed. A. N. Doane and Carol Braun Pasternack, 137–61. Madison: University of Wisconsin Press, 1991.

Norman, Diana. *Siena and the Virgin: Art and Politics in a Late-Medieval City State*. New Haven: Yale University Press, 2003.

Ornato, Ezio, ed. *La carta occidentale nel tardo Medioevo*. Rome: Istituto centrale per la patologia del libro, 2001.

Pagden, Anthony. "Dispossessing the Barbarian: The Language of Spanish Thomism and the Debate over Property Rights of American Indians." In *The Language of Political Theory in Early Modern Europe*, ed. Anthony Pagden, 79–98. Cambridge: Cambridge University Press, 1986.

Panofsky, Erwin. *Early Netherlandish Painting: Its Origins and Character*. New York: Harper and Row, 1971.

———. *The Life and Art of Albrecht Dürer*. Princeton: Princeton University Press, 1955.

Parkhurst, Charles P. "Madonna of the Writing Christ Child." *Art Bulletin* 23, no. 4 (1941): 292–306.

Parshall, Peter W. "Art and the Theater of Knowledge: The Origins of Print Collecting in Northern Europe." *Harvard University Art Museums Bulletin* 2, no. 3 (1994): 7–36.

———. "The Print Collection of Ferdinand, Archduke of Tyrol." *Jahrbuch der kunsthistorischen Sammlungen in Wien* 78 (1982): 147–48.

Pedersen, Johannes. "Writing Materials." In *The Arabic Book*, ed. Robert Hillenbrand; trans. Geoffrey French, 54–71. Princeton: Princeton University Press, 1984.

Perkinson, Stephen. "Engin and Artifice: Describing Creative Agency at the Court of France, ca. 1400." *Gesta* 41, no. 1 (2002): 51–67.

———. "Rethinking the Origins of Portraiture." *Gesta* 46, no. 2 (2007): 140–41.

Peter the Venerable. *Against the Inveterate Obduracy of the Jews*. Trans. Irven M. Resnick. Washington, D.C.: Catholic University of America Press, 2013.

Peters, Charles M., Joshua Rosenthal, and Teodile Urbina. "Otomi Bark Paper in Mexico: Commercialization of a Pre-Hispanic Technology." *Economic Botany* 41, no. 3 (1987): 423–32.

Petrarch, Francesco. *The Sonnets of Petrarch in the Original Italian Together with English Translations*. New York: Heritage, 1966.

Piccard, Gerhard. "Carta bombycina, carta papyri, pergamena greca: Ein Beitrag zur Geschichte der Beschreibstoffe im Mittelalter." *Archivalische Zeitschrift* (1965): 46–75.

Pincus, Debra. "Mark Gets the Message: Mantegna and the 'Praedestinatio' in Fifteenth-Century Venice." *Artibus et Historiae* 18, no. 35 (1997): 135–46.

Plano, Sara Cunchillos. "Inventario analítico de documentos de la series 'Cartas reales' de Jaime I del Archivo de la Corona de Aragón." *Jaime I y su época, 3, 4, y 5, X Congreso de Historia de la Corona de Aragón* 3 (1979): 485–508.

Plato. *Plato's Theaetetus*. Trans. David Bostock. Oxford: Clarendon, 1988.

Pliny the Elder. *Naturalis historia: English and Latin*. Vol. 4. Trans. H. Rackham. Cambridge, Mass.: Harvard University Press; London: Heinemann, 1940–63.

Podgorny, Irina. "Toward a Bureaucratic History of Archaeology: A Preliminary Essay." In *Historiographical Approaches to Past Archaeological Research*, ed. Gisela Eberhardt and Fabian Link, 47–67. Berlin: Topoi, 2015.

Popham, A. E., and Philip Pouncey. *Italian Drawings in the Department of Prints and Drawings in the British Museum*. London: Published by the Trustees of the British Museum, 1950.

Prazniak, Roxann. "Siena on the Silk Roads: Ambrogio Lorenzetti and the Mongol Global Century, 1250–1350." *Journal of World History* 21, no. 2 (2010): 177–217.

Prideaux, Humphrey. *The Old and New Testament Connected in the History of the Jews and Neigh-*

boring Nations, from the Declension of the Kingdoms of Israel and Judah to the Time of Christ. London: Printed for R. Knaplock, 1716–18.
Purchas, Samuel. *Purchas His Pilgrimes: In Five Books.* London: Printed by William Standsby, 1625.
Rappaport, Joanne, and Tom Cummins. *Beyond the Lettered City: Indigenous Literacies in the Andes.* Durham, N.C.: Duke University Press, 2012.
Reynolds, L. D. "The Elder Pliny." In *Texts and Transmission: A Survey of the Latin Classics,* 307–16. Oxford: Clarendon, 1984.
Reynolds, Terry S. *Stronger than a Hundred Men: A History of the Vertical Water Wheel.* Baltimore: Johns Hopkins University Press, 1983.
Rice, Yael. "Lines of Perception: European Prints and the Mughal kitābkhāna." In *Prints in Translation, 1450–1750: Image, Materiality, Space,* ed. Suzanne Karr Schmidt and Edward H. Wouk, 203–24. London: Routledge, 2017.
Richards, Gertrude Randolph Bramlette. *Florentine Merchants in the Age of the Medici.* Cambridge, Mass.: Harvard University Press, 1932.
Richardson, Brian. *Printing, Writers, and Readers in Renaissance Italy.* Cambridge: Cambridge University Press, 1998.
Rischel, Anna-Grethe. "Analysis of the Papermaker's Choice of Fibrous Materials and Technology along the Paper Road." In *Paper as a Medium of Cultural Heritage: Archaeology and Conservation 26th Congress–International Association of Paper Historians (Rome–Verona, August 30th–September 6th, 2002),* ed. Rosella Graziaplena, 232–8. Rome: Istituto centrale per la patologia del libro, 2004.
Robertson, Donald. *Mexican Manuscript Painting of the Early Colonial Period: The Metropolitan Schools.* New Haven: Yale University Press, 1959.
Robison, Andrew. *Paper in Prints.* Washington, D.C.: National Gallery of Art, 1977.
Robison, Andrew, and Klaus Albrecht Schröder, eds. *Albrecht Dürer: Master Drawings, Watercolors, and Prints from the Albertina.* Washington, D.C.: National Gallery of Art; Munich: Delmonico/Prestel, 2013.
Roman, L. "A History of Lost Tablets." *Classical Antiquity* 25, no. 2 (2006): 351–88.
Roper, Geoffrey. "Printed in Europe, Consumed in Ottoman Lands: European Books in the Middle East, 1514–1842." In *Books in Motion in Early Modern Europe,* ed. Daniel Bellingradt, Paul Nelles, and Jeroen Salman, 267–88. London: Palgrave, 2017.
Rorty, Richard. *Philosophy and the Mirror of Nature.* Princeton: Princeton University Press, 1979.
Rosand, David. *Drawing Acts: Studies in Graphic Expression and Representation.* Cambridge: Cambridge University Press, 2002.
Rouse, Richard H., and Mary A. Rouse. "The Vocabulary of Wax Tablets." In *Vocabulaire du livre et de l'ecriture au Moyen Âge: actes de la table ronde, Paris, 24–26 septembre 1987,* 220–30. Turnhout, Belg.: Brepols, 1989.
Röver-Kann, Anne. *Dürer-Zeit: die Geschichte der Dürer-Sammlung in der Kunsthalle Bremen.* Munich: Hirmer, 2012.
Russo, Alessandra. *The Untranslatable Image: A Mestizo History of the Arts in New Spain, 1500–1600.* Austin: University of Texas Press, 2014.
Sahagún, Bernardino de. *General History of the Things of New Spain: Florentine Codex.* Trans. Arthur J. O. Anderson and Charles E. Dibble. Santa Fe: School of American Research, 1950–82.
Salmon, Claudine. "La mission de Théodose de Lagrené et les enquêtes sur les textiles d'Insulide (1844–1846)." *Archipel* 75 (2008): 167–97.
Sander, Jochen, ed. *Albrecht Dürer: His Art in Context.* Munich: Prestel, 2013.
Sandstrom, Alan R., and Pamela Effrein Sandstrom. *Traditional Papermaking and Paper Cult Figures of Mexico.* Norman: University of Oklahoma Press, 1986.
Schäfer, Dagmar. *The Crafting of 10,000 Things: Knowledge and Technology in Seventeenth-Century China.* Chicago: University of Chicago Press, 2011.
Scheller, Robert W. *Exemplum: Model-Book Drawings and the Practice of Artistic Transmission in the Middle Ages (ca. 900–ca. 1470).* Trans. Michael Hoyle. Amsterdam: Amsterdam University Press, 1995.
———. "Reviewed Work(s): Das Skizzenbuch des Jaques Daliwe; Kommentar zur Faksimile Ausgabe des Liber picturatus A 74 der Staatsbibliothek

Berlin/DDR by Ulrike Jenni." *Simiolus: Netherlands Quarterly for the History of Art* 19, no. 3 (1989): 206–8.

Schenck, Kimberly. "Drawings under Scrutiny: The Materials and Techniques of Metalpoint." In *Drawing in Silver and Gold: Leonardo to Jasper Johns,* 9–24. Princeton: Princeton University Press, 2015.

Schlosser, Julius von. "Vademecum eines fahrenden Gesellen: Zur Kenntnis der künstlerischen Überlieferung im späten Mittelalter." *Jahrbuch der Kunsthistorischen Sammlungen des Allerhöchsten Kaiserhauses* 23 (1902): 279–338.

Schrader, Stephanie, ed. *Rembrandt and the Inspiration of India.* Los Angeles: Getty Research Institute, 2018.

Schröder, Klaus Albrecht, and Maria Luise Sternath. *Albrecht Dürer.* Vienna: Albertina; Ostfildern-Ruit, Ger.: Hatje Cantz, 2003.

Schroeder, Susan. "Writing Two Cultures: The Meaning of 'Amoxtli' (*Book*) in Nahua New Spain." In *New World, First Nations,* ed. David Cahill and Blanca Tovías, 13–35. Brighton: Sussex Academic Press, 2006.

Shanks, Torrey. *Authority Figures: Rhetoric and Experience in John Locke's Philosophical Thought.* University Park: Pennsylvania State University Press, 2014.

Siegert, Bernhard. *Cultural Techniques: Grids, Filters, Doors, and Other Articulations of the Real.* Trans. Geoffrey Winthrop-Young. New York: Fordham University Press, 2015.

———. *Passagiere und Papiere: Schreibakte auf der Schwelle zwischen Spanien und Amerika.* Munich: Wilhelm Fink, 2006.

Silver, Larry. *Peasant Scenes and Landscapes: The Rise of Pictorial Genres in the Antwerp Art Market.* Philadelphia: University of Pennsylvania Press, 2012.

Silver, Larry, and Jeffrey Chipps Smith, eds. *The Essential Dürer.* Philadelphia: University of Pennsylvania Press, 2010.

Simpson, Lesley Bird. *Exploitation of Land in Central Mexico in the Sixteenth Century.* Berkeley: University of California Press, 1952.

Singh, Kavita. *Mughal Birds in Imagined Gardens: Mughal Painting between Persia and Europe.* Los Angeles: Getty Research Institute, 2017.

Smith, Alistair. "'Germania' and 'Italia': Albrecht Dürer and Venetian Art." *Journal of the Royal Society of Arts* 127 (1979): 273–90.

Smith, Jeffrey Chipps. "Albrecht Dürer as Collector." *Renaissance Quarterly* 64, no. 1 (2011): 1–49.

Smith, Mary Elizabeth. "Why the Second Codex Selden Was Painted." In *Caciques and Their People: A Volume in Honor of Ronald Spores: Anthropological Papers* 89 (1994): 111–41.

Soll, Jacob. "From Note-Taking to Data Banks." *Intellectual History Review* 20, no. 3 (2010): 355–75.

Song Yingxing. *T'ien-kung k'ai-wu: Chinese Technology in the Seventeenth Century.* University Park: Pennsylvania State University Press, 1966.

Stallybrass, Peter, Roger Chartier, John Franklin Mowery, and Heather Wolfe. "Hamlet's Tables and Technologies of Writing in Renaissance England." *Shakespeare Quarterly* 55, no. 4 (2004): 379–419.

Starr, Frederick. *In Indian Mexico: A Narrative of Travel and Labor.* Chicago: Forbes, 1908.

Stevenson, Allan H. "Briquet and the Future of Paper Studies." In *Briquet's Opuscula: The Complete Works of Dr. C. M. Briquet without Les filigranes,* xv–1. Hilversum, Neth.: Paper Publications Society, 1955.

Stohr, Tomás. "Watermarks in Venezuelan Documents from Colonial Times." In *Paper as a Medium of Cultural Heritage: Archaeology and Conservation: 26th Congress IPH: Rome–Verona, August 30th–September 6th, 2002,* ed. Rosella Graziaplena, 209–22. Rome: Istituto centrale per la patologia del libro, 2004.

Strauss, Walter L. *The Complete Drawings of Albrecht Dürer.* New York: Abaris, 1974.

Stromer, Wolfgang Freiherr von. "Innovation und Wachstum im Spätmittelalter: Die Erfindung der Drahtmühle als Stimulator." *Tehnikgeschichte* 44 (1977): 89–120.

Te Heesen, Anke. "The Notebook: A Paper-Technology." In *Making Things Public,* ed. Bruno Latour and Peter Weibel, 582–89. Cambridge: Cambridge University Press, 2005.

Tietze, Hans. *Dürer als Zeichner und Aquarellist.* Vienna: A. Schroll, 1951.

Timmer, David E. "Providence and Perdition: Fray Diego de Landa Justifies His Inquisition against the Yucatecan Maya." *Church History* 66, no. 3 (1997): 477–88.

Tolnay, Charles de. *History and Technique of Old Master Drawings: A Handbook*. New York: Hacker, 1972.

Tsuen-Hsuin, Tsien. *Chemistry and Chemical Technology: Paper and Printing*. Vol. 5, pt. 1, in *Science and Civilization in China,* ed. Joseph Needham. Cambridge: Cambridge University Press, 1954.

———. *Written on Bamboo and Silk: The Beginnings of Chinese Books and Inscriptions*. Chicago: University of Chicago Press, 2004.

Tully, James. *A Discourse on Property: John Locke and His Adversaries*. Cambridge: Cambridge University Press, 1980.

Valle, Perla P. *Memorial de los indios de Tepetlaóztoc ó códice Kingsborough*. Mexico City: Instituto Nacional de Antropología e Historia, 1992.

Valls i Subirà, Oriol. *La historia del papel en España, siglos X–XIV*. 3 vols. Madrid: Empresa Nacional de Celulosas, 1978–82.

———. *Paper and Watermarks in Catalonia*. Ed. and trans. J. S. G. Simmons and B. J. van Ginneken-van de Kasteele. Amsterdam: Paper Publications Society, 1970.

Vasari, Giorgio. *La vita di Michelangelo, nelle redazione del 1550 e del 1568*. Ed. with commentary by Paola Barocchi. Milan: Riccardo Ricciardi, 1962.

Venuti, Lawrence. "Genealogies of Translation Theory: Jerome." *Boundary* 2 37, no. 3 (2010): 5–28.

Verma, Som Prakash. "Artists' Signatures in Miniatures of the Mughal School." In *Interpreting Mughal Painting: Essays on Art, Society, and Culture,* 28–43. Oxford: Oxford University Press, 2009.

Vida, Giorgio Levi Della. *Ricerche sulla formazione del più antico fondo dei manoscritti orientali della Biblioteca vaticana*. Vatican City: Biblioteca apostolica vaticana, 1939.

Vismann, Cornelia. *Files: Law and Media Technology*. Trans. Geoffrey Winthrop-Young. Stanford: Stanford University Press, 2008.

Voelkle, William. "Two New Drawings for the Boxwood Sketchbook in the Pierpont Morgan Library." *Gesta* 20 (1981): 243–45.

Voorn, Henk. "A Brief History of the Sizing of Paper." *Papermaker* 30, no. 1 (1961): 47–53.

Wakefield, Colin. "Arabic Manuscripts in the Bodleian Library: The Seventeenth-Century Collections." In *The "Arabick" Interests of Natural Philosophers in Seventeenth-Century England,* ed. G. A. Russell, 128–46. Leiden, Neth.: Brill, 1994.

Waley, Daniel Philip. *Siena and the Sienese in the Thirteenth Century*. Cambridge: Cambridge University Press, 1991.

Walsh, Paul. "The Medieval Merchant's Mark and Its Survival in Galway." *Journal of the Galway Archaeological and Historical Society* 45 (1993): 1–28.

Walsham, Alexandra. "The Social History of the Archive: Record-Keeping in Early Modern Europe." *Past and Present* 11 (2016): 9–48.

Weigel, Christopher. *Abbildung der gemein-nützlichen Haupt-Stände*. Regensburg, Ger.: n.p., 1698.

Weiß, Wisso. *Zeittafel zur Papiergeschichte*. Leipzig: Fachbuchverlag, 1983.

Whistler, Catherine. *Venice and Drawing, 1500–1800*. New Haven: Yale University Press, 2016.

Wiesner, Julius. "Die Faijûmer und Uschûmeiner Papiere, eine naturwissenschaftliche, mit Rücksicht auf die Erkennung alter und moderner Papiere auf die Entwicklung der Papierbereitung durchgeführte Untersuchung." *Mitteilungen aus der Sammlungen der Papyrus Erzherzog Rainer* 2, no. 3 (1887): 179–260.

Williams, Megan. "Unfolding Diplomatic Paper and Paper Practices in Early Modern Chancellery Archives." In *Praktiken der frühen Neuzeit: Akteure, Handlungen, Artefakte,* ed. Arndt Brendecke, 496–508. Cologne: Böhlau, 2015.

Williams, Sarah Jan. "An Author's Role in Fourteenth-Century Book Production: Guillaume de Machaut's 'livre où je met toutes mes choses.'" *Romania* 90 (1969): 433–54.

Winkler, Friedrich. *Die Zeichnungen Albrecht Dürers*. Berlin: Deutscher verein für kunstwissenschaft, 1936–37.

Wixom, William D. "Enthroned Madonna with the Writing Christ Child." *Bulletin of the Cleveland Museum of Art* 57, no. 9 (1970): 287–302.

Wölfflin, Heinrich. *Drawings of Albrecht Dürer*. New York: Dover, 1970.

Wood, Christopher S. *Forgery, Replica, Fiction: Temporalities of German Renaissance Art.* Chicago: University of Chicago Press, 2008.

Wood, Neal. "Tabula Rasa, Social Environmentalism, and the 'English Paradigm.'" *Journal of the History of Ideas* 53, no. 4 (1992): 647–68.

Wray, Shona Kelly. *Communities and Crisis: Bologna during the Black Death.* Leiden, Neth.: Brill, 2009.

Wroth, Lawrence C. *Some Reflections on the Books Arts in Early Mexico.* Cambridge, Mass.: Department of Printing and the Graphic Arts, Harvard University and Harvard College Library, 1945.

Yale, Elizabeth. "The History of Archives: The State of the Discipline." *Book History* 18 (2015): 332–59.

Zumthor, Paul. *Introduction à la poésie orale.* Paris: Seuil, 1983.

ILLUSTRATION CREDITS

The photographers and the sources of visual material other than those indicated in the captions are as follows. Every effort has been made to credit the photographers and the sources; if there are errors or omissions, please contact Yale University Press so that corrections can be made in any subsequent edition.

KHM-Museumsverband, Vienna (figs. 1, 2, 6, 7, 65)
Image courtesy of the Clark Art Institute, Williamstown, Massachusetts, USA (figs. 3, 4, 69, 70, 79, 80)
*93HR-2014, Houghton Library, Harvard University (figs. 5, 19, 20)
© British Library Board/Robana/Art Resource, NY (fig. 8)
Drawn by Bill Nelson (fig. 9)
Library of the Benedictine Abbey of Saint Paul's in Sankt Paul im Lavanttal (fig. 10)
© The Trustees of the British Museum, London (figs. 11, 12, 66, 71, 89, 100–102, 105, 110)
GEN B 5945 510*, Houghton Library, Harvard University (figs. 13, 14)
Ernst Kirchner Collection, Deutsches Museum, Munich (figs. 15–17)
bpk Bildagentur/Staatliche Museen/Jörg P. Anders/Art Resource, NY (fig. 18)
f Jpn40.1.2, Houghton Library, Harvard University (figs. 21, 22)
CHINOIS-5563, © BnF, Dist. RMN-Grand Palais (figs. 23–26)
RESERVE OE-112-PET FOL, © BnF, Dist. RMN-Grand Palais (figs. 27–29)
© Bodleian Library, University of Oxford [2019], Ms. Laud Or. 139 (fig. 30)
© Bodleian Library, University of Oxford [2019], Ms. Greaves 25 (fig. 31)
© Bodleian Library, University of Oxford [2019], Ms. Huntington 202 (fig. 32)
Copyright the Warburg Institute (fig. 33)
Library of Congress, Geography and Maps Division, Washington, D.C. (fig. 34)
Public Domain, Instituto Nacional de Antopología e Historia, Mexico City (fig. 35–38)
© Bodleian Library, University of Oxford, Ms. Arch. Selden. A. 1 (figs. 39, 40)
GEN WKR 14.2.8, Houghton Library, Harvard University (fig. 41)
Public Domain, Bibliothèque municipale de Lyon (figs. 42, 43)
Courtesy of Princeton University Library (fig. 46)
Scala/Art Resource, NY (figs. 47–50, 54, 96)
bpk Bildagentur/Rosgartenmuseum Konstanz/Art Resource, NY (fig. 51)
Universal Images Group/Art Resource, NY (fig. 52)
Album/Art Resource, NY (fig. 53)
Herzog August Bibliothek Wolfenbüttel: Cod. Guelf. 61.2 Aug. 8° (fig. 55)
Staatsbibliothek zu Berlin–Preussischer Kulturbesitz, Manuscript Department, Libr. pict. A 74, fol. Ib (fig. 56)
bpk Bildagentur/Staatsbibliothek zu Berlin, Stiftung Preussischer Kulturbesitz, Berlin, Germany/Art Resource, NY (figs. 57, 59)
Public Domain: The Cloisters Collection, 1954 (fig. 58)
The Morgan Library & Museum. MS M.346. Purchased by J. Pierpont Morgan (1837–1913), 1906 (figs. 60 [f.1v], 61–63, 64 [fol. 2r])
Erich Lessing/Art Resource (figs. 67, 77)
Museo Nacional Thyssen-Bornemisza/Scala/Art Resource (figs. 68, 78)
The Albertina Museum, Vienna (figs. 72–76, 85–87)

Rosenwald Collection, The National Gallery of Art, Washington, D.C. (figs. 81, 82)

Alisa Mellon Bruce Fund, the National Gallery of Art, Washington, D.C. (fig. 83)

Historisches Museum Frankfurt, Photo: Horst Ziegenfusz (fig. 84)

Robert Lehman Collection, 1975, The Metropolitan Museum of Art, New York (fig. 88)

The Rijksmuseum, Amsterdam (figs. 90, 91)

Royal Collection Trust/© Her Majesty Queen Elizabeth II 2018 (figs. 92, 93)

Mexicain 385, © BnF, Dist. RMN-Grand Palais (figs. 94, 95)

The Harkness Collection, Library of Congress, Washington, D.C. (figs. 97–99)

HIP/Art Resource, NY (fig. 103)

The Rijksmuseum, Amsterdam, RP-P-OB-147 (fig. 104)

© Schloss Schönbrunn Kultur und Betriebsges.m.b.H./Digitalisat: Salon Iris (fig. 106)

The Rijksmuseum, Amsterdam, RP-P-1954–135 (fig. 107)

Louis V. Bell Fund, 1967, The Metropolitan Museum of Art, New York (fig. 108)

Harris Brisbane Dick Fund, 1927, The Metropolitan Museum of Art, New York (fig. 109)

The Rijksmuseum Amsterdam, RP-P-OB-611 (fig. 111)

The Rijksmuseum, Amsterdam, RP-P-1962–121 (fig. 112)

Harris Brisbane Dick Fund, 1927, The Metropolitan Museum of Art, New York (fig. 113)

INDEX